Reviews of Environmental Contamination and Toxicology

VOLUME 208

For other titles published in this series, go to
http://www.springer.com/series/398

Reviews of Environmental Contamination and Toxicology

Perfluorinated Alkylated Substances

Editor
David M. Whitacre

Volume Editor
Pim de Voogt

VOLUME 208

 Springer

Coordinating Board of Editors

ISSN 0179-5953
ISBN 978-1-4614-2676-9   ISBN 978-1-4419-6880-7 (eBook)
DOI 10.1007/978-1-4419-6880-7
Springer New York Dordrecht Heidelberg London

Springer is part of Springer Science+Business Media (www.springer.com)

Special Foreword

Anthropogenic perfluorinated alkylated substances (PFASs) have recently gained in socio-economic and scientific interest. PFASs constitute an emerging group of environmental contaminants that possess physico-chemical as well as toxicological properties different from those of other halogenated compounds. PFASs are generally persistent in the environment and are detected in most parts of the aquatic and terrestrial ecosystems over a broad concentration range. PFASs have been found around the globe in the blood of the general population, as well as in a suite of wildlife, including those even from pristine and remote areas.

The findings of perfluorinated acids (PFAs) in remote areas, notably in the Arctic marine food chain, have spurred scientific research targeted at elucidating the origin and transport pathways of these substances. Two main sources for these contaminants have been proposed: atmospheric oxidation of volatile precursor compounds and long-range oceanic transport of directly emitted perfluorinated acids. Clear scientific evidence that favors one of these pathways over the other has not decisively surfaced in the scientific community thus far, partly because large knowledge gaps or uncertainties exist with regard to PFAS' physico-chemical properties, emission data, and environmental behavior. Analogously, knowledge gaps exist about the pathways and the extent of human exposure, hampering risk assessment and establishment of trustworthy guideline values for ambient air, food and drinking water, or occupational exposure.

The four review chapters that comprise this volume address several of the key knowledge gaps in chemistry and exposure to these PFASs. The emphasis of each of these chapters is summarized in the next few paragraphs.

The two major classes of PFAs, perfluorocarboxylic acids (PFCAs) and perfluorosulfonic acids (PFSAs), are both likely to be ionized at environmental pH, suggesting they will primarily be present in the aqueous phase. Long-range transport through water occurs slowly, on the order of decades. The ubiquitous environmental distribution of PFAs suggests a faster, atmospheric dissemination mechanism. The chapter by Young and Mabury reviews and addresses the known atmospheric formation mechanisms of PFAs, the chemistry of compounds that can form PFAs, atmospheric levels of precursors, and the significance of atmospheric formation of PFAs.

The analytical chemistry of PFAS is complicated by the presence of impurities and isomers. These result from PFAS manufacturing processes. The two most commonly used synthetic methods have produced products with very different isomeric purities. As a consequence, both branched and linear PFAs and PFA-precursor isomers exist in the environment. The extent to which branching patterns affect the physical, chemical, and biological properties of perfluorinated substances is of increasing scientific interest. It is hypothesized that branching patterns may affect properties such as environmental transport and degradation, partitioning, bioaccumulation, pharmacokinetics, and toxicity. It may even influence total PFA quantification, thus perhaps leading to questions about the accuracy of existing human and environmental exposure assessments. The chapter by Benskin et al. in this volume addresses the measurement and interpretation of isomer signatures in the environment.

Biodegradation has been proven to contribute significantly to the amounts of PFASs detected globally, especially for perfluorooctanoic acid. The Organization of Economic Co-operation and Development (OECD) has published a list containing more than 600 chemicals that may degrade into PFCAs. A small number of these chemicals have actually been shown to contribute to environmental levels of the PFCAs. The contribution from Frömel and Knepper to this volume reviews our understanding of the metabolism of PFASs.

The diet and inhalation of house dust are considered to be the most important exposure pathways of perfluorinated compounds for humans. Due to the widespread distribution, environmental degradation, and metabolism of the PFASs released into the environment, a very complex exposure situation exists. As a result, the relative contribution to human exposure from different routes or from a single source (e.g., diet) is not yet known. More specifically, it is currently unknown as to what extent exposure to drinking water, food, or dust contributes to the PFASs that are measured in human breast milk and blood. Data on levels of PFASs in the human diet are rather scarce. Food processing such as cooking and packaging may alter the concentration of PFASs in food and as a consequence affect the risk for humans. The contribution from D'Hollander et al. reviews the available information on routes of human exposure.

The initial idea to compose a special issue in the *Reviews* arose during an international workshop on Fluorinated Surfactants, held in Idstein, Germany in June 2008. Several of the contributions presented at that meeting reviewed the state of the PFAS science, and when I approached some of the presenters to probe their interest in contributing to a Special Issue of the *Reviews*, I received a spontaneous and positive response. The contributors to the actual issue in front of you are experts in some key areas of PFAS research, including atmospheric and reaction chemistry, analytical chemistry and toxicology, biosynthesis and biotransformation, food chemistry, and environmental chemistry. All authors have thoroughly updated their contributions from earlier versions, as the area of PFAS research is one of the most active in environmental chemistry. The importance of the understanding of the chemistry and toxicology of PFAS is underlined by international legislation such as the EC directive on the marketing and use of perfluorooctane sulfonate (PFOS) in products, the inclusion of PFOS on the persistent organic pollutants' (POPs) list of the Stockholm Convention, and the Canadian regulations regarding PFOS and its salts, by the large dietary exposure surveys running in North America, Japan, and Europe, and by the Stewardship program agreement between PFAS manufacturers and the United States Environmental Protection Agency (USEPA).

April 2010   *Pim de Voogt*

Foreword

International concern in scientific, industrial, and governmental communities over traces of xenobiotics in foods and in both abiotic and biotic environments has justified the present triumvirate of specialized publications in this field: comprehensive reviews, rapidly published research papers and progress reports, and archival documentations. These three international publications are integrated and scheduled to provide the coherency essential for non-duplicative and current progress in a field as dynamic and complex as environmental contamination and toxicology. This series is reserved exclusively for the diversified literature on "toxic" chemicals in our food, our feeds, our homes, recreational and working surroundings, our domestic animals, our wildlife, and ourselves. Tremendous efforts worldwide have been mobilized to evaluate the nature, presence, magnitude, fate, and toxicology of the chemicals loosed upon the Earth. Among the sequelae of this broad new emphasis is an undeniable need for an articulated set of authoritative publications, where one can find the latest important world literature produced by these emerging areas of science together with documentation of pertinent ancillary legislation.

Research directors and legislative or administrative advisers do not have the time to scan the escalating number of technical publications that may contain articles important to current responsibility. Rather, these individuals need the background provided by detailed reviews and the assurance that the latest information is made available to them, all with minimal literature searching. Similarly, the scientist assigned or attracted to a new problem is required to glean all literature pertinent to the task, to publish new developments or important new experimental details quickly, to inform others of findings that might alter their own efforts, and eventually to publish all his/her supporting data and conclusions for archival purposes.

In the fields of environmental contamination and toxicology, the sum of these concerns and responsibilities is decisively addressed by the uniform, encompassing, and timely publication format of the Springer triumvirate:

Reviews of Environmental Contamination and Toxicology [Vols. 1–97
(1962–1986) as Residue Reviews] for detailed review articles concerned
with any aspects of chemical contaminants, including pesticides, in the total
environment with toxicological considerations and consequences.

Bulletin of Environmental Contamination and Toxicology (Vol. 1 in 1966) for
   rapid publication of short reports of significant advances and discoveries in
   the fields of air, soil, water, and food contamination and pollution as well as
   methodology and other disciplines concerned with the introduction,
   presence, and effects of toxicants in the total environment.

Archives of Environmental Contamination and Toxicology (Vol. 1 in 1973) for
   important complete articles emphasizing and describing original
   experimental or theoretical research work pertaining to the scientific aspects
   of chemical contaminants in the environment.

Manuscripts for *Reviews* and the *Archives* are in identical formats and are peer
reviewed by scientists in the field for adequacy and value; manuscripts for the
Bulletin are also reviewed, but are published by photo-offset from camera-ready
copy to provide the latest results with minimum delay. The individual editors of
these three publications comprise the joint Coordinating Board of Editors with refer-
ral within the board of manuscripts submitted to one publication but deemed by
major emphasis or length more suitable for one of the others.

Coordinating Board of Editors

Preface

The role of *Reviews* is to publish detailed scientific review articles on all aspects of environmental contamination and associated toxicological consequences. Such articles facilitate the often complex task of accessing and interpreting cogent scientific data within the confines of one or more closely related research fields.

In the nearly 50 years since *Reviews of Environmental Contamination and Toxicology* (formerly *Residue Reviews*) was first published, the number, scope, and complexity of environmental pollution incidents have grown unabated. During this entire period, the emphasis has been on publishing articles that address the presence and toxicity of environmental contaminants. New research is published each year on a myriad of environmental pollution issues facing people worldwide. This fact, and the routine discovery and reporting of new environmental contamination cases, creates an increasingly important function for *Reviews*.

The staggering volume of scientific literature demands remedy by which data can be synthesized and made available to readers in an abridged form. *Reviews* addresses this need and provides detailed reviews worldwide to key scientists and science or policy administrators, whether employed by government, universities, or the private sector.

There is a panoply of environmental issues and concerns on which many scientists have focused their research in past years. The scope of this list is quite broad, encompassing environmental events globally that affect marine and terrestrial ecosystems; biotic and abiotic environments; impacts on plants, humans, and wildlife; and pollutants, both chemical and radioactive; as well as the ravages of environmental disease in virtually all environmental media (soil, water, air). New or enhanced safety and environmental concerns have emerged in the last decade to be added to incidents covered by the media, studied by scientists, and addressed by governmental and private institutions. Among these are events so striking that they are creating a paradigm shift. Two in particular are at the center of ever-increasing media as well as scientific attention: bioterrorism and global warming. Unfortunately, these very worrisome issues are now superimposed on the already extensive list of ongoing environmental challenges.

The ultimate role of publishing scientific research is to enhance understanding of the environment in ways that allow the public to be better informed. The term "informed public" as used by Thomas Jefferson in the age of enlightenment conveyed the thought of soundness and good judgment. In the modern sense, being "well informed" has the narrower meaning of having access to sufficient information. Because the public still gets most of its information on science and technology from TV news and reports, the role for scientists as interpreters and brokers of scientific information to the public will grow rather than diminish. Environmentalism is the newest global political force, resulting in the emergence of multinational consortia to control pollution and the evolution of the environmental ethic. Will the new politics of the twenty-first century involve a consortium of technologists and environmentalists, or a progressive confrontation? These matters are of genuine concern to governmental agencies and legislative bodies around the world.

For those who make the decisions about how our planet is managed, there is an ongoing need for continual surveillance and intelligent controls to avoid endangering the environment, public health, and wildlife. Ensuring safety-in-use of the many chemicals involved in our highly industrialized culture is a dynamic challenge, for the old, established materials are continually being displaced by newly developed molecules more acceptable to federal and state regulatory agencies, public health officials, and environmentalists.

Reviews publishes synoptic articles designed to treat the presence, fate, and, if possible, the safety of xenobiotics in any segment of the environment. These reviews can be either general or specific, but properly lie in the domains of analytical chemistry and its methodology, biochemistry, human and animal medicine, legislation, pharmacology, physiology, toxicology, and regulation. Certain affairs in food technology concerned specifically with pesticide and other food-additive problems may also be appropriate.

Because manuscripts are published in the order in which they are received in final form, it may seem that some important aspects have been neglected at times. However, these apparent omissions are recognized, and pertinent manuscripts are likely in preparation or planned. The field is so very large and the interests in it are so varied that the editor and the editorial board earnestly solicit authors and suggestions of underrepresented topics to make this international book series yet more useful and worthwhile.

Justification for the preparation of any review for this book series is that it deals with some aspect of the many real problems arising from the presence of foreign chemicals in our surroundings. Thus, manuscripts may encompass case studies from any country. Food additives, including pesticides, or their metabolites that may persist into human food and animal feeds are within this scope. Additionally, chemical contamination in any manner of air, water, soil, or plant or animal life is within these objectives and their purview.

Manuscripts are often contributed by invitation. However, nominations for new topics or topics in areas that are rapidly advancing are welcome. Preliminary communication with the editor is recommended before volunteered review manuscripts are submitted.

Summerfield, NC, USA   *David M. Whitacre*

Contents

Contributors

Jonathan P. Benskin Division of Analytical and Environmental Toxicology, Department of Laboratory Medicine and Pathology, University of Alberta, Edmonton, AB T6G 2G3, Canada, jbenskin@ualberta.ca

Lieven Bervoets Laboratory for Ecophysiology, Biochemistry and Toxicology, Department of Biology, University of Antwerp, 2020 Antwerp, Belgium, lieven.bervoets@ua.ac.be

Wim De Coen Laboratory for Ecophysiology, Biochemistry and Toxicology, Department of Biology, University of Antwerp, 2020 Antwerp, Belgium, wim.decoen@ua.ac.be

Wendy D'Hollander Laboratory for Ecophysiology, Biochemistry and Toxicology, Department of Biology, University of Antwerp, 2020 Antwerp, Belgium, wendy.dhollander@ua.ac.be

Amila O. De Silva Environment Canada, Water Science and Technology Directorate, Burlington, ON L7R 4A6, Canada, amila.desilva@ec.gc.ca

Pim de Voogt Institute for Biodiversity and Ecosystem Dynamics, University of Amsterdam, POBOX 94240, 1090 GE, Amsterdam, The Netherlands; KWR Watercycle Research Institute, Nieuwegein, The Netherlands, w.p.devoogt@uva.nl

Tobias Frömel Institute for Analytic Research, Hochschule Fresenius, 65510 Idstein, Germany, froemel@hs-fresenius.de

Thomas P. Knepper Institute for Analytic Research, Hochschule Fresenius, 65510 Idstein, Germany, knepper@hs-fresenius.de

Scott A. Mabury Department of Chemistry, University of Toronto, Toronto, ON M5S 3H6, Canada, smabury@chem.utoronto.ca

Jonathan W. Martin Division of Analytical and Environmental Toxicology, Department of Laboratory Medicine and Pathology, University of Alberta, Edmonton, AB T6G 2G3, Canada, jon.martin@ualberta.ca

Cora J. Young Department of Chemistry, University of Toronto, Toronto, ON M5S 3H6, Canada, cora.young@noaa.gov

Atmospheric Perfluorinated Acid Precursors: Chemistry, Occurrence, and Impacts

Cora J. Young and Scott A. Mabury

Contents

S.A. Mabury (✉)
Department of Chemistry, University of Toronto, Toronto, ON M5S 3H6, Canada
e-mail: smabury@chem.utoronto.ca

P. de Voogt (ed.), *Reviews of Environmental Contamination and Toxicology*,
Reviews of Environmental Contamination and Toxicology 208,
DOI 10.1007/978-1-4419-6880-7_1, © Springer Science+Business Media, LLC 2010

1 Introduction

Interest in perfluorinated acids (PFAs) began over the past decade with the real-ization that PFAs were present in organisms (Giesy and Kannan 2001), including humans (Hansen et al. 2001; Taves 1968). Further studies indicated long-chain con-geners were bioaccumulative (Martin et al. 2003a; b) and could magnify within a food chain (Tomy et al. 2004). Global monitoring for PFAs revealed these com-pounds were found ubiquitously in ocean water (Scott et al. 2005; Yamashita et al. 2005), precipitation (Scott et al. 2006), biota (Houde et al. 2006) and human blood (Kannan et al. 2004).

The two major classes of perfluorinated acids (PFAs), perfluorocarboxylic acids (PFCAs) and perfluorosulfonic acids (PFSAs), are both likely to be ionized at envi-ronmental pH, suggesting they will be present primarily in the aqueous phase. Long-range transport through water occurs slowly, on the order of decades. The ubiquitous distribution of PFAs suggests a faster, atmospheric dissemination mecha-nism. In addition, not all PFAs observed in the environment have been commercially produced.

It is well established that PFCAs can be formed in the atmosphere from the hydrolysis of acyl halides from compounds such as hydrochlorofluorocarbons (HCFCs) and hydrofluorocarbons (HFCs). However, in the past 5 years, a num-ber of studies have examined the potential for other compounds to form acyl halides or PFCAs through more complex mechanisms. These mechanisms are of particular interest because of the potential to form bioaccumulative long-chain PFCAs. Recent work has also shown the potential for PFSAs to be formed through atmospheric oxidation of precursor compounds.

PFA precursors are exclusively anthropogenic products, manufactured primarily for two major categories of products. The first is the surfactant and fluoropolymer industry, which includes compounds that are used in the synthesis of water- and oil-repellent compounds. PFA precursors are also marketed as chlorofluorocarbon (CFC) replacements, for uses that include coolants, solvents, and fire suppressors. PFA precursors are found in other industries, though not as frequently. In Table 1, acronyms used in this chapter for PFAs and their precursors are defined.

Table 1 List of acronyms for the chemicals addressed in this review

Acronym	Name	Structure
PFA	Perfluorinated acid	N/A
PFCA	Perfluorocarboxylic acid	$CF_3(CF_2)_xC(O)OH$
TFA	Trifluoroacetic acid	$CF_3C(O)OH$
PFPrA	Perfluoropropionic acid	$CF_3CF_2C(O)OH$
PFBA	Perfluorobutanoic acid	$CF_3(CF_2)_2C(O)OH$
PFPeA	Perfluoropentanoic acid	$CF_3(CF_2)_3C(O)OH$
PFHxA	Perfluorohexanoic acid	$CF_3(CF_2)_4C(O)OH$
PFHeA	Perfluoroheptanoic acid	$CF_3(CF_2)_5C(O)OH$
PFOA	Perfluorooctanoic acid	$CF_3(CF_2)_6C(O)OH$
PFNA	Perfluorononanoic acid	$CF_3(CF_2)_7C(O)OH$
PFDA	Perfluorodecanoic acid	$CF_3(CF_2)_8C(O)OH$

Table 1 (continued)

Acronym	Name	Structure
PFUnA	Perfluoroundecanoic acid	$CF_3(CF_2)_9C(O)OH$
PFDoA	Perfluorododecanoic acid	$CF_3(CF_2)_{10}C(O)OH$
PFTrA	Perfluorotridecanoic acid	$CF_3(CF_2)_{11}C(O)OH$
PFTA	Perfluorotetradecanoic acid	$CF_3(CF_2)_{12}C(O)OH$
PFPA	Perfluoropentadecanoic acid	$CF_3(CF_2)_{13}C(O)OH$
PFSA	Perfluorosulfonic acid	$CF_3(CF_2)_xSO_3H$
PFBS	Perfluorobutanesulfonic acid	$CF_3(CF_2)_3SO_3H$
PFOS	Perfluorooctanesulfonic acid	$CF_3(CF_2)_7SO_3H$
HFO	Hydrofluoroolefin	$C_xF_yH_{(2x-y)}$
HFC	Hydrofluorocarbon	$C_xF_yH_{(2x+2-y)}$
HFC-134a	1,1,1,2-tetrafluoroethane	CF_3CH_2F
HFC-125	1H-Perfluoroethane	CF_3CF_2H
HFC-329	1H-Perfluoropropane	$CF_3(CF_2)_2H$
HFC-227	2H-Perfluoropropane	CF_3CHFCF_3
CFC	Chlorofluorocarbon	$C_xF_yCl_{(2x+2-y)}$
HCFC	Hydrochlorofluorocarbon	$C_xF_yCl_zH_{(2x+2-y-z)}$
HCFC-124	1-Chloro-1,2,2,2-tetrafluoroethane	$CF_3(CF_2)_xCHFCl$
HCFC-123	1,1-Dichloro-2,2,2-trifluoroethane	CF_3CHCl_2
HCFC-225ca	1,1-Dichloro-2,2,3,3,3-pentafluoropropane	$CF_3CF_2CHCl_2$
PFAL	Perfluorinated aldehyde	$CF_3(CF_2)_xC(O)H$
PFAL hydrate	Perfluorinated aldehyde hydrate	$CF_3(CF_2)_xCH(OH)_2$
FTAL	Fluorotelomer aldehyde	$CF_3(CF_2)_xCH_2C(O)H$
FTOH	Fluorotelomer alcohol	$CF_3(CF_2)_xCH_2CH_2OH$
oFTOH	Odd fluorotelomer alcohol	$CF_3(CF_2)_xCH_2OH$
FTO	Fluorotelomer olefin	$CF_3(CF_2)_xCH=CH_2$
FTI	Fluorotelomer iodide	$CF_3(CF_2)_xCH_2CH_2I$
FTAc	Fluorotelomer acrylate	$CF_3(CF_2)_xCH_2CH_2OC(O)$ $CH=CH_2$
FTCA	Fluorotelomer carboxylic acid	$CF_3(CF_2)_xCH_2C(O)OH$
NAFSA	N-Alkyl-perfluoroalkanesulfonamide	$CF_3(CF_2)_xSO_2N(H)C_yH_{(2y+1)}$
NAFBSA	N-Alkyl-perfluorobutanesulfonamide	$CF_3(CF_2)_3SO_2N(H)C_yH_{(2y+1)}$
NEtFBSA	N-Ethyl-perfluorobutanesulfonamide	$CF_3(CF_2)_3SO_2N(H)C_2H_5$
NAFOSA	N-Alkyl-perfluorooctanesulfonamide	$CF_3(CF_2)_7SO_2N(H)C_yH_{(2y+1)}$
PFOSA	Perfluorooctanesulfonamide	$CF_3(CF_2)_7SO_2NH_2$
NMeFOSA	N-Methyl-perfluorooctanesulfonamide	$CF_3(CF_2)_7SO_2N(H)CH_3$
NEtFOSA	N-Ethyl-perfluorooctanesulfonamide	$CF_3(CF_2)_7SO_2N(H)C_2H_5$
NAFSE	N-Alkyl-perfluoroalkanesulfamido ethanol	$CF_3(CF_2)_xSO_2N(C_yH_{(2y+1)})$ CH_2CH_2OH
NAFBSE	N-Alkyl-perfluorobutanesulfamido ethanol	$CF_3(CF_2)_3SO_2N(C_yH_{(2y+1)})$ CH_2CH_2OH
NMeFBSE	N-Methyl-perfluorobutanesulfamido ethanol	$CF_3(CF_2)_3SO_2N(CH_3)CH_2CH_2OH$
NEtFBSE	N-Ethyl-perfluorobutanesulfamido ethanol	$CF_3(CF_2)_3SO_2N(C_2H_5)CH_2CH_2OH$
NAFOSE	N-Alkyl-perfluorooctanesulfamido ethanol	$CF_3(CF_2)_7SO_2N(C_yH_{(2y+1)})$ CH_2CH_2OH
NMeFOSE	N-Methylperfluorooctanesulfamido ethanol	$CF_3(CF_2)_7SO_2N(CH_3)CH_2CH_2OH$
NEtFOSE	N-Ethyl-perfluorooctanesulfamido ethanol	$CF_3(CF_2)_7SO_2N(C_2H_5)CH_2CH_2OH$

In this review, we will address the known atmospheric formation mechanisms of PFAs, the chemistry of compounds that can form PFAs, atmospheric levels of precursors, and the significance of atmospheric formation of PFAs.

2 Mechanisms of Atmospheric Formation of Perfluorinated Acids

2.1 Perfluorocarboxylic Acids

2.1.1 Mechanisms for Atmospheric Formation of Perfluoroacyl Halides

Chemistry of Perfluoroacyl Halides

The atmospheric fate of perfluoroacyl fluorides can be limited by photolysis or hydrolysis:

$$CF_3(CF_2)_xC(O)F + h\nu \rightarrow \text{products} \tag{1}$$

$$CF_3(CF_2)_xC(O)F + H_2O_{(1)} \rightarrow \text{products} \tag{2}$$

Absorption cross-sections for $CF_3C(O)F$ have been measured and a quantum yield for reaction (1) of approximately unity was determined (Rattigan et al. 1993). Using this information, photolysis lifetimes of 3,000 years (Rattigan et al. 1993) and 7,500 years (Wild et al. 1996) were estimated. In contrast, hydrolysis of $CF_3C(O)F$ occurs on a faster timescale. Obtaining atmospherically relevant hydrolysis measurements is difficult (Wallington et al. 1994) and ones made have led to vastly differing results. However, Kanakidou et al. (1995) determined lifetimes for in-cloud loss by including time in the gas phase, probability of collision with a droplet, the sticking coefficient, and the actual hydrolysis rate using a model. Hydrolysis rates used in the model spanned more than two orders of magnitude and these encompassed measured values. Determined lifetimes ranged from 5.9 to 7.6 d, suggesting little sensitivity to the actual hydrolysis rate. A hydrolysis lifetime of 13.6 d has also been estimated by Wild et al. (1996). Although no studies have focused on the fate of longer-chained perfluoroacyl fluorides, it is expected that hydrolysis will still be a dominant factor for them as well. Increasing perfluorinated chain length has been observed to lead to a slight red-shift of the absorption spectrum, as well as an increase in absorption cross-section (Chiappero et al. 2006). These differences are unlikely to have a large impact on the photolytic lifetime. This suggests that the principal fate of $CF_3(CF_2)_xC(O)F$ is hydrolysis, which occurs with a lifetime on the order of days to weeks. The exclusive product of hydrolysis of perfluoroacyl fluorides is the corresponding PFCA.

Similarly, the fate of perfluoroacyl chlorides can be determined through photolysis and hydrolysis:

$$CF_3(CF_2)_xC(O)Cl + h\nu \rightarrow products \quad (3)$$

$$CF_3(CF_2)_xC(O)Cl + H_2O_{(l)} \rightarrow products \quad (4)$$

Rattigan et al. (1993) and Meller and Moortgat (1997) have measured the absorption cross-section of $CF_3C(O)Cl$ and also determined a quantum yield of dissociation of near unity. Utilizing these measurements, photolysis lifetimes of 33 d (Rattigan et al. 1993), 40–60 d (Meller and Moortgat 1997), <86 d (Hayman et al. 1994), and 56.4 d (Wild et al. 1996) were estimated. Kanakidou et al. (1995) used various hydrolysis rates to determine lifetimes with respect to in-cloud hydrolysis, which ranged from 5.4 to 6.9 d. A similar estimate of 11.1 d was made by Wild et al. (1996). Although hydrolysis proceeds faster than photolysis, photolytic degradation will likely play a role in the fate of perfluoroacyl chlorides. Analysis of $CF_3C(O)Cl$ in a two-dimensional model demonstrated that hydrolysis was the major fate (Hayman et al. 1994). Photolysis was observed to be less significant, with potential importance primarily in the upper troposphere or lower stratosphere, forming $C(O)F_2$, HCl, and HF. The degree to which photolysis is important may be dependent on perfluorinated chain length, as Chiappero et al. (2006) observed a slight red-shift on increasing from $-CF_3$ to a longer-perfluorinated chain, along with an increased absorption cross-section. For $CF_3C(O)Cl$, the fraction reacting via hydrolysis was estimated at 0.6 (Tang et al. 1998). The relative importance of these two processes has not been studied for other chain-length perfluoroacyl chlorides, but it is clear that both pathways will be important in the overall atmospheric fate. The sole product of hydrolysis of $CF_3(CF_2)_xC(O)Cl$ is the corresponding PFCA.

Mixed Halide Mechanism

Mixed halides of the structure $CF_3(CF_2)_xCHXY$, where X is fluorine or chlorine and Y is chlorine or bromine can form PFCAs. Abstraction of the hydrogen atom is followed by reaction with molecular oxygen and NO to yield an acyl radical. The radical decomposes, eliminating the Y atom and leaving a perfluoroacyl fluoride or chloride (Fig. 1). This has been observed to occur for CF_3CHFCl (Edney and Driscoll 1992), CF_3CHCl_2 (Edney et al. 1991), and $CF_3CHClBr$ (Bilde et al. 1998). Given that gas-phase radical mechanisms have been shown to be independent of perfluorinated chain length (Hurley et al. 2004b), this mechanism should apply to all volatile fluorinated chemicals of this class.

Perfluorinated Radical Mechanism

Perfluorinated radicals $(CF_3(CF_2)_xCF_2^\bullet$, where $x \geq 0)$ are formed from the atmospheric oxidation of many polyfluorinated and perfluorinated compounds. The most common environmental fate of these perfluorinated radicals is "unzipping" to yield a carbon-equivalent number of molecules of carbonyl fluoride (COF_2) (Wallington et al. 1994). However, a lower-yield degradation pathway initially described by Ellis et al. (2004) can lead to the formation of PFCAs (Fig. 2). The initial step in the

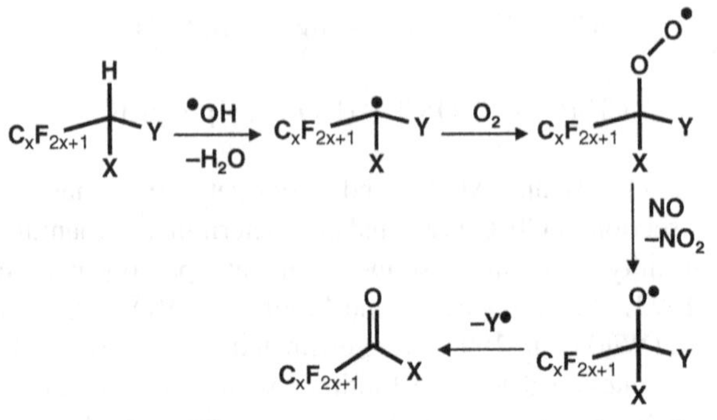

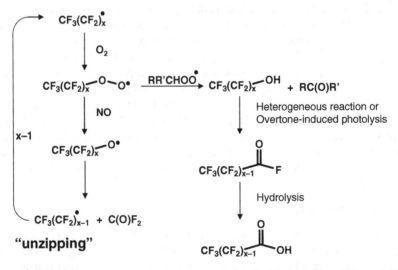

Fig. 1 Mechanism of perfluoroacyl fluoride formation from mixed halides, where X = F,Cl and Y = Cl,Br

Fig. 2 Mechanism of PFCA formation via perfluorinated radicals (adapted from Ellis et al. 2004)

atmospheric oxidation of perfluorinated radicals is reaction with oxygen to yield a perfluoroalkyl peroxy radical:

$$CF_3(CF_2)_xCF_2^\bullet + O_2 \rightarrow CF_3(CF_2)_xCF_2OO^\bullet \tag{5}$$

This radical has a number of possible fates. The most common is reaction with the abundant radical, nitric oxide (NO), to yield a perfluoro alkoxy radical, which loses COF_2 to yield a perfluorinated radical containing one fewer carbon atoms:

$$CF_3(CF_2)_xCF_2OO^\bullet + NO \rightarrow CF_3(CF_2)_xCF_2O^\bullet \tag{6a}$$

$$CF_3(CF_2)_xCF_2O^\bullet \rightarrow C(O)F_2 + CF_3(CF_2)_{x-1}CF_2^\bullet \tag{7}$$

The cycling of reactions (5), (6a), and (7) characterizes the unzipping pathway that dominates the fate of perfluorinated radicals under typical atmospheric conditions. In remote areas where levels of NO are lower, reaction with peroxy radicals (ROO$^\bullet$) may become important. The reaction of a perfluoro peroxy radical with ROO$^\bullet$ may yield a perfluoro alkoxy radical that goes on to react via reaction (7) to form COF_2:

$$CF_3(CF_2)_xCF_2OO^\bullet + ROO^\bullet \rightarrow CF_3(CF_2)_xCF_2O^\bullet + RO^\bullet + O_2 \qquad (6b)$$

If the ROO$^\bullet$ contains an alpha hydrogen (RR$'$CHOO$^\bullet$), the perfluoro peroxy radical can react to yield a perfluorinated alcohol:

$$CF_3(CF_2)_xCF_2OO^\bullet + RR'CHCOO^\bullet \rightarrow CF_3(CF_2)_xCF_2OH + R'C(O)R \qquad (6c)$$

Perfluorinated alcohols are inherently unstable and quickly lose HF to yield perfluoroacyl fluorides by a mechanism that has not been entirely elucidated:

$$CF_3(CF_2)_xCF_2OH \rightarrow CF_3(CF_2)_xC(O)F + HF \qquad (8)$$

Suggested mechanisms for the loss of HF from perfluorinated alcohols include heterogeneous reaction (Schneider et al. 1996) and overtone-induced photolysis (Young and Donaldson 2007). As discussed above, perfluoroacyl fluorides readily hydrolyze, producing PFCAs.

$$CF_3(CF_2)_xC(O)F + H_2O \rightarrow CF_3(CF_2)_xC(O)OH \qquad (9)$$

Yields of PFCAs from perfluorinated radicals have not been directly studied, but will be related to the ratio of RR$'$CHOO$^\bullet$ to other reactive species. In urban areas, where NO is abundant, it is unlikely that reaction (6c) will be significant. However, in remote, low-NO_x environments, such as the Arctic and over the ocean, the production of PFCAs is likely to occur more readily (Wallington et al. 2006).

Although the impact of perfluorinated chain length on the yield of PFCAs has not been explicitly determined, it has been shown that under NO_x-free conditions, perfluorinated radicals (($CF_3(CF_2)_xCF_2^\bullet$, $x = 7, 5, 3$)) form a homologous series of shorter-chain PFCAs (Ellis et al. 2004). This occurs as a result of the removal of COF_2 through the unzipping mechanism and the formation of perfluorinated radicals of decreasing chain lengths. Thus, from the perfluorinated radical, $CF_3CF_2^\bullet$, trifluoroacetic acid (TFA) will be formed, while perfluorinated radicals of the structure $CF_3(CF_2)_xCF_2^\bullet$, where $x \geq 1$, will form PFCAs ($CF_3(CF_2)_nC(O)OH$) of multiple chain lengths, from $n = x$ to TFA ($n = 0$).

Hydrofluoroolefin (HFO) Mechanism

More recently, formation of perfluoroacyl fluorides from fluorinated alkenes has been observed (Fig. 3). Polyfluorinated and perfluorinated propenes and a perfluorinated butene with a terminal double bond were shown to form molar yields

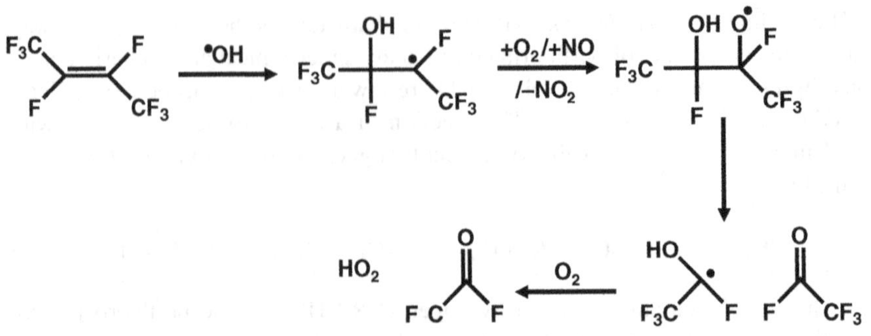

Fig. 3 Mechanism for the atmospheric oxidation of CF$_3$CF=CFCF$_3$ initiated by hydroxyl radicals (adapted from Young et al. 2009b)

indistinguishable from 100% of perfluoroacyl fluorides through reaction with both chlorine atoms and hydroxyl radicals (Acerboni et al. 2001; Hurley et al. 2007; Mashino et al. 2000; Young et al. 2009b). Similarly, perfluorobutene with an internal double bond was shown to produce CF$_3$C(O)F in a molar yield indistinguishable from 200% through reaction with chlorine and hydroxyl radicals (Young et al. 2009b). Yields for these reactions were unchanged in the presence and absence of NO$_x$. Because perfluorinated chain length has not been shown to affect reaction mechanism in experiments with alcohols (Hurley et al. 2004b) and acids (Hurley et al. 2004c), it is likely that any alkene, in which a C–F bond and a perfluorinated alkane chain appear on one side of the double bond, would be expected to follow this pathway, producing perfluoroacyl fluorides, and subsequently, PFCAs, in 100% molar yield. Fluorinated alkenes form PFCAs in 100 or 200% yield, under typical atmospheric conditions, in the presence or absence of NO$_x$.

2.1.2 Mechanisms for Direct Atmospheric Formation of PFCAs

Perfluoroacyl Peroxy Radical Mechanism

A second mechanism that has been observed to lead to formation of PFCAs is via reaction of perfluoroacyl peroxy radicals CF$_3$(CF$_2$)$_x$C(O)OO$^\bullet$, where $x \geq 0$) with HO$_2$ (Ellis et al. 2004). Acetyl peroxy radicals (CH$_3$C(O)OO$^\bullet$) are known to degrade by one of three channels, forming peracetic acid, acetic acid, or acetoxy radicals (Hasson et al. 2004):

$$CH_3C(O)OO^\bullet + HO_2 \rightarrow CH_3C(O)OOH + O_2 \qquad (10a)$$

$$CH_3C(O)OO^\bullet + HO_2 \rightarrow CH_3C(O)OH + O_3 \qquad (10b)$$

$$CH_3C(O)OO^\bullet + HO_2 \rightarrow CH_3C(O)O^\bullet + O_2 + OH \qquad (10c)$$

Studies with $CF_3(CF_2)_xC(O)OO^\bullet$ (where $x = 0–3$) demonstrated that perfluoroa-cyl peroxy radicals can undergo analogous reactions, yielding perfluorocarboxylic peracids, PFCAs, or perfluoroacyl oxy radicals (Sulbaek Andersen et al. 2003a, 2004b):

$$CF_3(CF_2)_xC(O)OO^\bullet + HO_2 \rightarrow CF_3(CF_2)_xC(O)OOH + O_2 \quad (11a)$$

$$CF_3(CF_2)_xC(O)OO^\bullet + HO_2 \rightarrow CF_3(CF_2)_xC(O)OH + O_3 \quad (11b)$$

$$CF_3(CF_2)_xC(O)OO^\bullet + HO_2 \rightarrow CF_3(CF_2)_xC(O)O^\bullet + O_2 + OH \quad (11c)$$

By analogy with the hydrocarbon reactions, these reactions are hypothesized to take place via a tetroxide intermediate (Fig. 4) (Moortgat et al. 1989), though this intermediate has never been observed experimentally. In general, a trend of increased PFCA and decreased radical yields with increasing chain length was observed. The dominant products were dependent on chain length (Table 2), with radical propagation by reaction (11c) dominating for $CF_3C(O)OO^\bullet$, with a yield of 0.52 ± 0.05 and PFCA formation by reaction (11b) dominating for $C_4F_9C(O)OO^\bullet$, with a yield of 0.73 ± 0.18 (Hurley et al. 2006). The product of

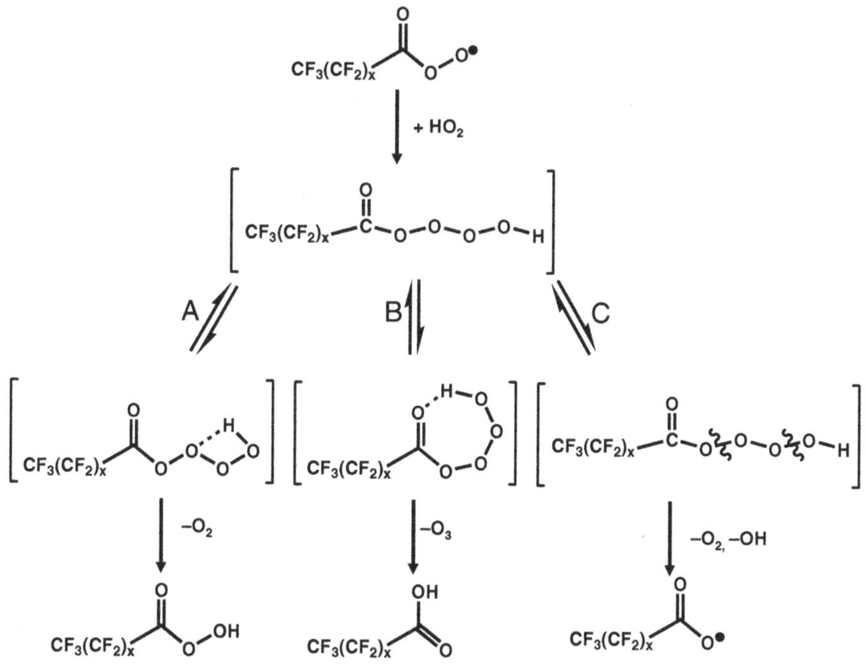

Fig. 4 Mechanism of PFCA formation via perfluoroacyl peroxy radicals (adapted from Sulbaek Andersen et al. 2003a)

Table 2 Product yields of reactions (10) and (11) (adapted from Hurley et al. 2006)

R	Product yields			References
	RC(O)OOH + O_2	RC(O)OH + O_3	RC(O)O$^\bullet$ + O_2 + OH	
CH_3	0.40 ± 0.16	0.20 ± 0.08	0.40 ± 0.16	Hasson et al. (2004)
CF_3	0.09 ± 0.04	0.39 ± 0.04	0.52 ± 0.05	Hurley et al. (2006)
CF_3CF_2	<0.12	0.50 ± 0.08	0.50 ± 0.08	Hurley et al. (2006)
$CF_3CF_2CF_2$	<0.16	0.53 ± 0.11	0.47 ± 0.11	Hurley et al. (2006)
$CF_3CF_2CF_2CF_2$	<0.27	0.73 ± 0.18	0.27 ± 0.18	Hurley et al. (2006)

reaction (11a), the perfluoroperacid, was detected only for $CF_3C(O)OO^\bullet$ (Sulbaek Andersen et al. 2004b). In reactions of longer-chain-length perfluoroacyl peroxy radicals, the corresponding perfluorocarboxylic peracid was below detection limits (Sulbaek Andersen et al. 2003a, 2004b). These product distributions are in contrast to those observed for analogous hydrocarbon reactions. Specifically, in observations of acetyl peroxy radicals, the peracetic acid is a more dominant product (Hasson et al. 2004). It must be cautioned that smog chamber experiments (Hurley et al. 2006; Sulbaek Andersen et al. 2003a, 2004b) cannot quantitatively describe the fraction of PFCAs formed by this reaction under environmental conditions. In smog chamber experiments, perfluoroacyl peroxy radicals were produced through reactions of the corresponding perfluorinated aldehydes (PFALs, $CF_3(CF_2)_xC(O)H$, $x = 1$–3) with chlorine radicals. The behavior of PFALs under smog chamber conditions is not well understood and could impact the formation of perfluoroacyl radicals from the PFAL. In addition, smog chamber experiments were performed at 296 ± 2 K, while common atmospheric temperatures range from 220 to 300 K. Studies of reaction (10) show negative temperature dependence, characteristic of a more stable intermediate complex at lower temperatures (Moortgat et al. 1989). It is possible that a longer-lived tetroxide intermediate would be more likely to undergo the complex rearrangement necessary to form acids under these conditions (pathway B, Fig. 4).

Finally, under atmospheric conditions, the reactions of NO_x with perfluoroacyl peroxy radicals will compete with the reactions described above:

$$CF_3(CF_2)_xC(O)OO^\bullet + NO \rightarrow CF_3(CF_2)_xC(O)O^\bullet + NO_2 \tag{11d}$$

$$CF_3(CF_2)_xC(O)OO^\bullet + NO_2 \rightarrow CF_3(CF_2)_xC(O)O_2NO_2 \tag{11e}$$

It is difficult to determine the fraction of perfluoroacyl peroxy radicals that will react with HO_2 in the atmosphere. In urban areas, levels of NO_x greatly exceed levels of HO_2, leading to reactions (11d) and (11e) that then dominate over the reaction with HO_2. Reactions with HO_2 are expected to be more prevalent in remote areas where HO_2 and NO_x levels are similar.

Perfluorinated Aldehyde (PFAL) Hydrate Mechanism

The formation of PFCAs was observed from the atmospheric oxidation of PFAL hydrates (Sulbaek Andersen et al. 2006). Reaction of $CF_3CH(OH)_2$ with chlorine atoms in the absence of NO_x was demonstrated to form TFA as the primary product in a yield indistinguishable from 100%.

$$CF_3CH(OH)_2 + Cl \rightarrow CF_3C(O)OH \tag{12}$$

The formation of $CF_3C(O)OH$ was also observed through reaction with hydroxyl radicals, with a yield close to 100%, though this was not independently quantified. This reaction is expected to occur via rearrangement of the peroxy radical as shown in Fig. 5. The production of PFCAs from PFAL hydrates in the presence of NO_x has not been investigated. Mechanisms of gas-phase reactions have been observed to be independent of perfluorinated chain length (Hurley et al. 2004b), suggesting that PFAL hydrates of all chain lengths will form PFCAs.

Fig. 5 Proposed mechanism for the formation of PFCAs from PFAL hydrates

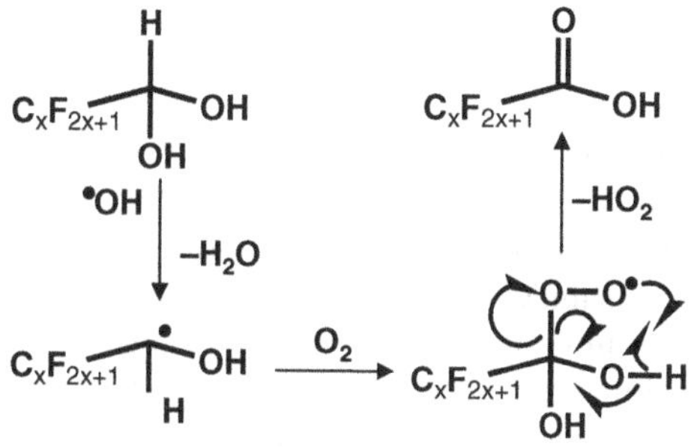

2.2 Perfluorosulfonic Acids (PFSAs)

PFSAs, $(CF_3(CF_2)_xSO_3H)$, can be formed from atmospheric reactions of perfluorosulfonamides. Perfluorobutane sulfonate (PFBS) was observed from the atmospheric oxidation of N-methyl perfluorobutane sulfonamidoethanol (NMeFBSE, $C_4F_9SO_2N(CH_3)CH_2CH_2OH$) (D'eon et al. 2006). The proposed mechanism to account for this observation is shown in Fig. 6 and involves the addition of hydroxyl radicals to the sulfone double bond. This results in a sulfonyl radical that can cleave the S–N bond to yield PFBS and a nitrogen-centered radical. Alternatively, the C–S bond can be broken, which yields a sulfamic acid and a perfluorinated radical that subsequently can decompose into PFCAs as discussed in Section *Perfluorinated Radical Mechanism*. The ratio of PFBS to PFCAs formed from NMeFBSE was approximately 1:10, suggesting higher probability of C–S bond breakage. The degradation of NMeFBSE is the sole observation of a PFSA

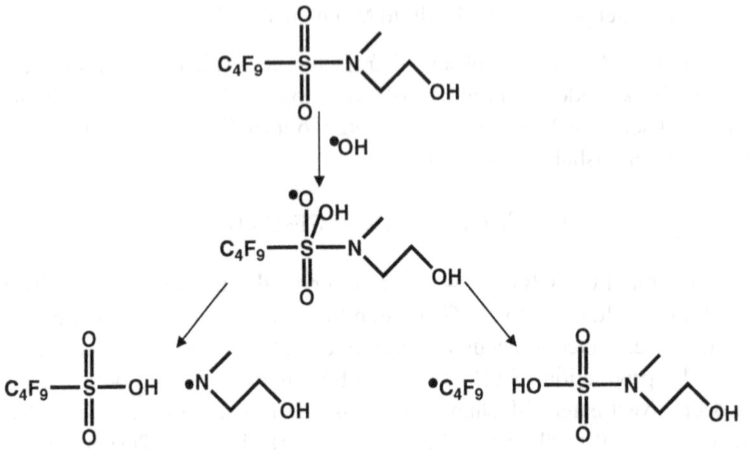

Fig. 6 Proposed mechanism of PFBA formation from NMeFBSE (adapted from D'eon et al. 2006)

being formed from atmospheric oxidation. The class of perfluorosulfonamides is diverse, with multiple perfluorinated chain lengths and substitutions on the amine group. It is possible to extrapolate the results from NMeFBSE to N-methylperfluoro-sulfonamides with different perfluorinated chain lengths, as this has been shown to have little effect on atmospheric chemistry (Hurley et al. 2004b, c). However, it is difficult to speculate on how substitution at the amine might affect the possible production of PFSAs from the atmospheric oxidation of perfluorosulfonamides.

3 Chemistry of Perfluorinated Acid (PFA) Precursors

A simplified scheme depicting the chemistry of perfluorinated acid precursors is shown in Fig. 7, and each of these precursors is discussed in detail below. It should be noted that for many of the described compounds, chlorine radicals have been used as a proxy to understand reaction mechanisms for hydroxyl radicals, because they can be produced easily and cleanly in an experimental setup. Chlorine radicals may be important locally, but concentrations are usually too low to impact the overall fate of organic compounds. However, they can shed light on hydroxyl radical mechanisms. In particular, when the mechanism for reaction with a hydroxyl radical is hydrogen abstraction, chlorine atom reactions provide a reasonable proxy. Some concern has been raised about stability of radical products following the reaction of chlorine atoms as compared to hydroxyl radicals (Waterland and Dobbs 2007); nonetheless, in most cases, chlorine atoms remain a useful tool for understanding atmospheric degradation pathways for organic compounds.

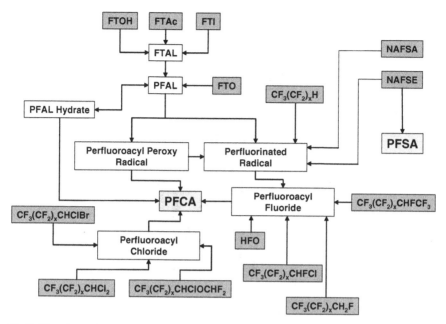

Fig. 7 Transformation pathways of perfluorinated acid precursors. *Gray-shaded* compound names are commercially produced

3.1 Volatile Anesthetics

3.1.1 $CF_3(CF_2)_x CHClBr$

Atmospheric Lifetime

The atmospheric fate of a single product in this class has been characterized. The commercial product, $CF_3CHClBr$, marketed as an anesthetic under the name Halothane, can be degraded by reaction with atmospheric oxidants:

$$CF_3CHClBr + Cl \rightarrow products \qquad (13)$$

$$CF_3CHClBr + OH \rightarrow products \qquad (14)$$

The kinetics of reaction (13) have been studied using relative rate methods (Table 3), yielding a rate constant of k (Cl + $CF_3CHClBr$) = $(2.0 \pm 0.3) \times 10^{-14}$ cm^3/molecule/s (Bilde et al. 1998). The low concentrations of chlorine atoms in the atmosphere (Singh et al. 1996) preclude reaction (13) from being of atmospheric significance for $CF_3CHClBr$. Of more atmospheric relevance is reaction (14), which has been examined by a few research groups (Table 3) (Brown et al.

Table 3 Summary of chlorine atom- and hydroxyl radical-initiated kinetics of volatile anesthetics

	Structure	x	Rate constant (cm^3/molecule/s)	T (K)	Method	References
Cl	$CF_3(CF_2)_xCHCIBr$	0	$(2.0 \pm 0.3) \times 10^{-14}$	296	Relative rate	Bilde et al. (1998)
	$CF_3(CF_2)_xCHClOCHF_2$	0	$(5.4 \pm 0.5) \times 10^{-15}$	295	Relative rate	Wallington et al. (2002)
OH	$CF_3(CF_2)_xCHCIBr$	0	$(6.0 \pm 0.4) \times 10^{-14}$	303	Discharge flow–resonance fluorescence	Brown et al. (1990)
		0	$(4.5 \pm 0.4) \times 10^{-14}$	298	Discharge flow–electron paramagnetic resonance spectrometry	Orkin and Khamaganov (1993)
		0	$(1.5 \pm 0.3) \times 10^{-14}$	298	Laser long-path absorption	Langbein et al. (1999)
	$CF_3(CF_2)_xCHClOCHF_2$	0	$(2.1 \pm 0.7) \times 10^{-14}$	298	Discharge flow–resonance fluorescence	Brown et al. (1990)
		0	$(1.7 \pm 0.3) \times 10^{-14}$	298	Laser long-path absorption	Langbein et al. (1999)

1990; Langbein et al. 1999; Orkin and Khamaganov 1993). There is significant discrepancy among the measured values, and a preferred value has been cited as 4.6×10^{-14} cm^3/molecule/s (Atkinson et al. 2008), based on the rate determined by Orkin and Khamaganov (1993). An atmospheric lifetime based on this value was estimated as 1.2 years, by comparison with CH$_3$CCl$_3$ (Orkin and Khamaganov 1993).

The atmospheric lifetime of CF$_3$CHClBr can also be limited by photolysis:

$$CF_3CHClBr + h\nu \rightarrow products \tag{15}$$

This reaction has been the subject of two studies, both of which observed photolysis of CF$_3$CHClBr to occur slowly in the troposphere. A tropospheric photolytic lifetime was determined by Bilde et al. (1998) as 5–10 years and by Langbein et al. (1999) as 105 years. Thus, the atmospheric fate of CF$_3$CHClBr will be dominated by reaction with hydroxyl radicals, with a lifetime of approximately 1.2 years. Perfluorinated chain length does not impact atmospheric kinetics (Hurley et al. 2004b, c), suggesting that the lifetime will be the same for all compounds of this class.

Products of Atmospheric Degradation

The products of reaction (13) have been measured as a proxy for the more relevant atmospheric reaction with hydroxyl radicals (reaction (14)) (Bilde et al. 1998). The sole product of the reaction was observed to be CF$_3$C(O)Cl, with a yield close to unity. This mechanism was determined to occur via the pathway described in *Mixed Halide Mechanism* Section and confirms CF$_3$(CF$_2$)$_x$CHClBr as a class of PFCA precursors.

3.1.2 CF$_3$(CF$_2$)$_x$CHClOCHF$_2$

Atmospheric Lifetime

The atmospheric fate of a single compound of this class (CF$_3$CHClOCHF$_2$), which is marketed as an anesthetic under the trade name Isoflurane, has been studied with respect to reaction with atmospheric oxidants:

$$CF_3CHClOCHF_2 + Cl \rightarrow products \tag{16}$$

$$CF_3CHClOCHF_2 + OH \rightarrow products \tag{17}$$

The kinetics of reaction (16) have been the subject of a single study (Table 3) in which a reaction occurred that had a rate of k (Cl + CF$_3$CHClOCHF$_2$) = $(5.4 \pm 0.5) \times 10^{-15}$ cm^3/molecule/s (Wallington et al. 2002). Chlorine atoms are not present in the atmosphere in high enough quantities to significantly impact the environmental fate of CF$_3$CHClOCHF$_2$ (Singh et al. 1996). Reaction (17) has been measured directly in two studies, which are in agreement within experimental error

(Table 3) (Brown et al. 1990; Langbein et al. 1999). A rate constant range can be derived from these measurements as k (OH + CF$_3$CHClOCHF$_2$) = 1.7–2.1 × 10^{-15} cm^3/molecule/s. These rates will likely also apply to longer-perfluorinated-chain congeners, given the independence of kinetics to this molecular property (Hurley et al. 2004b, c). Lifetimes were estimated with respect to hydroxyl radical reaction in the troposphere as 3.1 years (Brown et al. 1990) and 5.9 years (Langbein et al. 1999). Photolysis of CF$_3$CHClOCHF$_2$ was also shown to be insignificant in the troposphere (Langbein et al. 1999), so the atmospheric fate of this compound is determined by reaction with hydroxyl radical in the troposphere that has a lifetime on the order of a few years.

Products of Atmospheric Degradation

The products of reaction (16) have been studied as a proxy for reaction (17), which is the dominant fate of CF$_3$CHClOCHF$_2$ (Wallington et al. 2002). The reaction can be initiated by hydrogen abstraction in two ways:

$$CF_3CHClOCHF_2 + Cl/OH \rightarrow CF_3C(^\bullet)ClOCHF_2 + HCl/H_2O \qquad (18a)$$

$$CF_3CHClOCHF_2 + Cl/OH \rightarrow CF_3CHClOC(^\bullet)F_2 + HCl/H_2O \qquad (18b)$$

Experiments have demonstrated that reaction (18a) dominated in chlorine-initiated reactions, occurring 95% of the time (Wallington et al. 2002). The fate of the radical product of reaction (18a) is reaction with oxygen and NO to form an alkoxy radical that can decompose to form CF$_3$C(O)OCHF$_2$. The dominant atmospheric fate of this compound is hydrolysis to form TFA. It is likely that this mechanism is independent of perfluorinated chain length, and that all compounds of the structure CF$_3$(CF$_2$)$_x$CHClOCHF$_2$ will react to form CF$_3$(CF$_2$)$_x$C(O)OCHF$_2$ and thus, PFCAs.

3.2 Hydrochlorofluorocarbons (HCFCs)

3.2.1 CF$_3$(CF$_2$)$_x$CHFCl

Atmospheric Lifetime

The sole commercial product of the general structure CF$_3$(CF$_2$)$_x$CHFCl is HCFC-124 (CF$_3$CHFCl). Kinetics for reactions of this compound in the atmosphere have been well established. The primary tropospheric fate of HCFC-124 is reaction with hydroxyl radical:

$$CF_3CHFCl + OH \rightarrow products \qquad (19)$$

Kinetics of reaction (19) have been the subject of five studies, using both absolute and relative methods (Table 4) (Gierczak et al. 1991; Howard and Evenson 1976; Hsu and DeMore 1995; Watson et al. 1979; Yamada et al. 2000). There are considerable differences among the measurements. A value of k (OH + CF_3CHFCl) = 8.7×10^{-15} cm^3/molecule/s has been recommended by IUPAC (Atkinson et al. 2008) after considering the reliability of all measurements and is based on the values of Watson et al. (1979), Gierczak et al. (1991), and Hsu and DeMore (1995). The kinetics of other congeners of this compound class are likely similar, as perfluorinated chain length does not typically impact kinetics (Hurley et al. 2004b, c). Tropospheric lifetimes determined by reaction with hydroxyl radical have been estimated between 6.0 and 7.6 years using different modeling techniques (Kanakidou et al. 1995; Orlando et al. 1991; Prather and Spivakovsky 1990). This long tropospheric lifetime indicates that transport to the stratosphere and reactions that occur there could be important in determining the overall atmospheric fate of CF_3CHFCl. A lifetime of 87 years for loss of CF_3CHFCl to the stratosphere has been determined using a three-dimensional model (Kanakidou et al. 1995). In the stratosphere, HCFCs can undergo photolysis:

$$CF_3CHFCl + hv \rightarrow products \tag{20}$$

The absorbance characteristics of CF_3CHFCl have been studied over wavelengths and temperatures relevant in the stratosphere (Gillotay and Simon 1991; Orlando et al. 1991). Photolysis was not observed to be a significant stratospheric sink; rather, reactions with hydroxyl radical or $O(^1D)$ are more likely important (Orlando et al. 1991).

$$CF_3CHFCl + O(^1D) \rightarrow products \tag{21}$$

Kinetics of reaction (21) have been the subject of a single study that determined a rate of k ($CF_3CHFCl + O(^1D)$) = 8.6×10^{-11} cm^3/molecule/s (Warren et al. 1991).

An overall atmospheric lifetime of $6.1-8.0$ years was determined (Houghton et al. 2001; Orlando et al. 1991), which is similar to the lifetime determined by reaction with hydroxyl radicals in the troposphere, indicating stratospheric reactions are slow and have a minor impact on the atmospheric fate of CF_3CHFCl.

Products of Atmospheric Degradation

The products of reaction (19) have been characterized, using chlorine atoms as a surrogate for hydroxyl radicals (Edney and Driscoll 1992; Tuazon and Atkinson 1993a). These studies produced $CF_3C(O)F$ as the sole product (see *Mixed Halide Mechanism* Section):

$$CF_3CHFCl + Cl/OH \rightarrow CF_3C(O)F + HCl/H_2O \tag{22}$$

Table 4 Summary of chlorine atom- and hydroxyl radical-initiated kinetics of hydrochlorofluorocarbons (HCFCs)

	Structure	x	Rate constant (cm^3/molecule/s)	T (K)	Method	References
Cl	$CF_3(CF_2)_xCHCl_2$	0	$(1.22 \pm 0.18) \times 10^{-14}$	295	Relative rate	Wallington and Hurley (1992)
	$CF_3(CF_2)_xCHFCl$	0	$(1.24 \pm 0.19) \times 10^{-14}$	296	Discharge flow–laser magnetic resonance	Howard and Evenson (1976)
		0	$(9.4 \pm 0.3) \times 10^{-15}$	301	Flash photolysis–resonance fluorescence	Watson et al. (1979)
		0	$(9.44 \pm 0.75) \times 10^{-15}$	298	Discharge flow–laser magnetic resonance/flash photolysis–laser-induced fluorescence	Gierczak et al. (1991)
		0	7.67×10^{-15}	298	Relative rate (averaged)	Hsu and DeMore (1995)
		0	$(1.08 \pm 0.16) \times 10^{-14}$	297	Pulsed laser photolysis–laser-induced fluorescence	Yamada et al. (2000)
OH	$CF_3(CF_2)_xCHCl_2$	0	$(2.84 \pm 0.43) \times 10^{-14}$	296	Discharge flow–laser magnetic resonance	Howard and Evenson (1976)
		0	$(3.6 \pm 0.4) \times 10^{-14}$	298	Flash photolysis–resonance fluorescence	Watson et al. (1979)
		0	$(3.86 \pm 0.19) \times 10^{-14}$	293	Discharge flow–resonance fluorescence	Clyne and Holt (1979)
		0	$(3.52 \pm 0.28) \times 10^{-14}$	298	Flash photolysis–resonance fluorescence	Liu et al. (1990)
		0	$(5.9 \pm 0.6) \times 10^{-14}$	303	Discharge flow–resonance fluorescence	Brown et al. (1990)
		0	$(3.69 \pm 0.37) \times 10^{-14}$	298	Discharge flow–laser magnetic resonance/flash photolysis–laser-induced fluorescence	Gierczak et al. (1991)

Table 4 (continued)

Structure	x	Rate constant (cm^3/molecule/s)	T (K)	Method	References
	0	3.38×10^{-14}	298	Relative rate	Hsu and DeMore (1995)
	0	$(3.67 \pm 0.24) \times 10^{-14}$	296.3	Pulsed laser photolysis–laser-induced fluorescence	Yamada et al. (2000)
	1	$(3.7 \pm 0.8) \times 10^{-14}$	300	Discharge flow–resonance fluorescence	Brown et al. (1990)
	1	$(2.60 \pm 0.29) \times 10^{-14}$	298	Flash photolysis–resonance fluorescence	Zhang et al. (1991)
	1	$(2.41 \pm 0.24) \times 10^{-14}$	295	Discharge flow–laser-induced fluorescence	Nelson et al. (1992)

There has also been some study of products of reactions (20). Reaction of CF_3CHFCl with $O(^1D)$ has been demonstrated to result in quenching of the oxygen atom 31% of the time.

$$CF_3CHFCl + O^1(D) \rightarrow CF_3CHFCl + O(^3P) \tag{23}$$

Other products of this reaction have not been characterized, but this reaction is expected to be insignificant in the overall fate of CF_3CHFCl. Rather, reaction (22) is the likely fate of CF_3CHFCl and is expected to occur according to the mechanism described in *Mixed Halide Mechanism* Section and to be independent of chain length. Thus, the atmospheric fate of $CF_3(CF_2)_xCHFCl$ can be described by reaction with hydroxyl radicals in the troposphere and formation of $CF_3(CF_2)_xC(O)F$, and subsequently, PFCAs, in 100% yield.

3.2.2 $CF_3(CF_2)_xCHCl_2$

Atmospheric Lifetime

Two HCFCs of the structure $CF_3(CF_2)_xCHCl_2$ have been commercially produced: HCFC-123 ($x = 0$) and HCFC-225ca ($x = 1$). Studies have examined the kinetics of reaction with chlorine atoms and hydroxyl radicals:

$$CF_3(CF_2)_xCHCl_2 + Cl \rightarrow products \tag{24}$$

$$CF_3(CF_2)_xCHCl_2 + OH \rightarrow products \tag{25}$$

The kinetics of reaction (24) have been measured using relative rate methods for CF_3CHCl_2. A rate constant of k (Cl + CF_3CHCl_2) = $(1.22 \pm 0.18) \times 10^{-14}$ cm^3/molecules/s was determined (Wallington and Hurley 1992). Concentrations of chlorine atoms in the environment are insufficient to impact the atmospheric fate of these compounds (Singh et al. 1996). The kinetics of reaction (25) have been the subject of numerous studies (Table 4) (Brown et al. 1990; Clyne and Holt 1979; Gierczak et al. 1991; Howard and Evenson 1976; Hsu and DeMore 1995; Liu et al. 1990; Nelson Jr. et al. 1992; Watson et al. 1979; Yamada et al. 2000; Zhang et al. 1991). A preferred value of reaction (25) for CF_3CHCl_2 has been given by IUPAC as 3.6×10^{-14} cm^3/molecule/s (Atkinson et al. 2008), based on the values of Howard and Evenson (1976), Watson et al. (1979), Liu et al. (1990), Gierczak et al. (1991), Yamada et al. (2000), and Hsu and DeMore (1995). A preferred value is also given by IUPAC for reaction (25) for $CF_3CF_2CHCl_2$ of 3.6×10^{-14} cm^3/molecule/s (Atkinson et al. 2008), based on the work of Zhang et al. (1991) and Nelson et al. (1992). From the similarity of the recommended values and those listed in Table 4, it does not appear that there is a chain-length effect for reaction of $CF_3(CF_2)_xCHCl_2$ with hydroxyl radicals.

Tropospheric lifetimes with respect to reaction with hydroxyl radical have been estimated to range from 1.2 to 1.55 years using a variety of modeling techniques

(Kanakidou et al. 1995; Prather and Spivakovsky 1990; Wild et al. 1996). The reaction of $CF_3(CF_2)_xCHCl_2$ with hydroxyl radicals could also occur within the stratosphere. Wild et al. (1996) estimated a stratospheric lifetime for CF_3CHCl_2 with respect to reaction OH of 8.84 years.

Photolysis of $CF_3(CF_2)_xCHCl_2$ through scission of the C−Cl bond could also be an atmospheric loss process:

$$CF_3(CF_2)_xCHCl + h\nu \rightarrow \text{products} \qquad (26)$$

Absorption cross-sections have been measured for both CF_3CHCl_2 (Gillotay and Simon 1991; Nayak et al. 1996; Orlando et al. 1991) and $CF_3CF_2CHCl_2$ (Braun et al. 1991). Absorption cross-sections for $CF_3CF_2CHCl_2$ were observed to be slightly higher than those for CF_3CHCl_2 (Braun et al. 1991), consistent with known effects of increasing perfluorinated chain length. The tropospheric photolytic lifetime of CF_3CHCl_2 was determined as 2.7×10^5 years by Wild et al. (1996) using a two-dimensional model. Photolysis is much more likely to occur in the stratosphere, where available wavelengths can excite the C–Cl chromophore. In this region, it is suggested that photolysis is the dominant fate of $CF_3(CF_2)_xCHCl_2$ (Braun et al. 1991), with a lifetime of 55.7 years (Wild et al. 1996).

Reaction with $O(^1D)$ is another potential fate of $CF_3(CF_2)_xCHCl_2$ in the stratosphere:

$$CF_3(CF_2)_xCHCl_2 + O(^1D) \rightarrow \text{products} \qquad (27)$$

The rate for reaction of CF_3CHCl_2 as $k (O(^1D) + CF_3CHCl_2) = 3.6 \times 10^{-14} cm^3/molecule/s$ (Warren et al. 1991) has been determined in a single study. The stratospheric lifetime for this process has been estimated, using a two-dimensional model, as 334 years (Wild et al. 1996), and thus is of lesser significance than photolysis.

Overall atmospheric lifetimes of 1.4–1.73 and 2.1 years were determined for CF_3CHCl_2 (Houghton et al. 2001; Orlando et al. 1991; Wild et al. 1996) and $CF_3CF_2CHCl_2$ (Houghton et al. 2001), respectively. Total tropospheric lifetime estimates for CF_3CHCl_2 range from 1.2 to 1.6 years (Kanakidou et al. 1995; Orlando et al. 1991; Prather and Spivakovsky 1990). The similarity of these estimated lifetimes suggests that the overall atmospheric fate of $CF_3(CF_2)_xCHCl_2$ is dominated by reaction with hydroxyl radicals in the troposphere.

Products of Atmospheric Degradation

The products of tropospheric oxidation of CF_3CHCl_2 have been thoroughly examined (Edney et al. 1991; Hayman et al. 1994; Tuazon and Atkinson 1993a), where reactions initiated by chlorine atoms were used as a proxy for hydroxyl radical-initiated reactions. Initial reaction of CF_3CHCl_2 with chlorine atoms yields a radical $(CF_3C(^\bullet)Cl_2)$ that goes on to react with molecular oxygen and NO to yield an alkoxy radical $(CF_3C(O^\bullet)Cl_2)$. This peroxy radical has two possible fates (Hayman et al. 1994):

$$CF_3C(O^\bullet)Cl_2 \rightarrow CF_3 + C(O)Cl_2 \qquad (28a)$$

$$CF_3C(O^\bullet)Cl_2 \rightarrow CF_3C(O)Cl + Cl \qquad (28b)$$

Edney et al. (1991), Tuazon and Atkinson (1993a), and Hayman et al. (1994) all observed complete conversion to $CF_3C(O)Cl$, indicating reaction (28a) to be of minor importance in the atmosphere and confirming the importance of the mechanism described in *Mixed Halide Mechanism* Section.

There has also been some study of products of reaction (27). Reaction of CF_3CHCl_2 with $O(^1D)$ has been demonstrated to result in quenching of the oxygen atom 21% of the time.

$$CF_3CHCl_2 + O(^1D) \rightarrow CF_3CHCl_2 + O(^3P) \qquad (29)$$

Although the products of reaction (27) have not been examined in other studies, this reaction is predicted to be of little importance compared to the tropospheric reaction of $CF_3(CF_2)_xCHCl_2$ with hydroxyl radicals, which will form $CF_3(CF_2)_xC(O)Cl$ and, subsequently, PFCAs.

3.3 Hydrofluorocarbons (HFCs, Non-telomer Based)

3.3.1 Saturated Hydrofluorocarbons (HFCs)

$CF_3(CF_2)_xCH_2F$

Atmospheric Lifetime

The primary commercial product of this class, and the sole compound of atmospheric study, is HFC-134a (CF_3CH_2F). As one of the most commonly used HFCs, the atmospheric fate of CF_3CH_2F has been widely examined. This compound can react with chlorine atoms or hydroxyl radicals:

$$CF_3CH_2F + Cl \rightarrow products \qquad (30)$$

$$CF_3CH_2F + OH \rightarrow products \qquad (31)$$

Kinetics of reaction (30) have been studied using direct and relative rate methods (Table 5) (Louis et al. 1997; Sawerysyn et al. 1992; Tuazon et al. 1992; Wallington and Hurley 1992). A preferred value of reaction (30) has been determined as 1.5×10^{-15} cm^3/molecule/s (Atkinson et al. 2008), based on the values of Tuazon et al. (1992), Wallington and Hurley (1992), and Louis et al. (1997). Global concentrations of chlorine atoms are very low and are unlikely to impact

Table 5 Summary of kinetics of chlorine atom- and hydroxyl radical-initiated reactions of saturated hydrofluorocarbons (HFCs)

	Structure	x	Rate constant (cm^3/molecule/s)	Method	References
Cl	$CF_3(CF_2)_xCH_2F$	0	$(1.6 \pm 0.3) \times 10^{-15}$	Discharge flow–mass spectrometry	Sawerysyn et al. (1992)
		0	$(1.38 \pm 0.18) \times 10^{-15}$	Relative rate	Wallington and Hurley (1992)
		0	$(1.6 \pm 0.3) \times 10^{-15}$	Relative rate	Tuazon et al. (1992)
	$CF_3(CF_2)_xCHF_2$	0	$(1.4 \pm 0.3) \times 10^{-15}$	Discharge flow–mass spectrometry	Louis et al. (1997)
		0	$(2.4 \pm 0.5) \times 10^{-16}$	Relative rate	Tuazon et al. (1992)
		0	$(2.6 \pm 0.6) \times 10^{-16}$	Relative rate	Sehested et al. (1993)
		0	$(2.93 \pm 0.23) \times 10^{-16}$	Relative rate	Young et al. (2009a)
		2	$(2.77 \pm 0.20) \times 10^{-16}$	Relative rate	Young et al. (2009a)
	$CF_3(CF_2)_xCHFCF_3$	0	$(3.8) \times 10^{-17}$	Laser photolysis–laser-induced fluorescence	Zellner et al. (1994)
OH	$CF_3(CF_2)_xCH_2F$	0	$(4.5 \pm 1.2) \times 10^{-17}$	Relative rate	Møgelberg et al. (1996)
		0	$(5.5 \pm 0.7) \times 10^{-15}$	Discharge flow–resonance fluorescence	Clyne and Holt (1979)
		0	$(5.2 \pm 0.6) \times 10^{-15}$	Flash photolysis–resonance absorption	Martin and Paraskevopoulos (1983)
		0	$(8.44 \pm 0.73) \times 10^{-15}$	Discharge flow–resonance fluorescence	Jeong et al. (1984)
		0	6.25×10^{-15}	Discharge flow–resonance fluorescence	Brown et al. (1990)
		0	$(5.18 \pm 0.70) \times 10^{-15}$	Flash photolysis–resonance fluorescence	Liu et al. (1990)
		0	$(4.34 \pm 0.35) \times 10^{-15}$	Discharge flow–laser magnetic resonance/flash photolysis–laser-induced fluorescence	Gierczak et al. (1991)

Table 5 (continued)

Structure	x	Rate constant (cm^3/molecule/s)	T (K)	Method	References
$CF_3(CF_2)_xCH_2F$	0	$(2.38 \pm 0.22) \times 10^{-15}$	270	Discharge flow–resonance fluorescence	Zhang et al. (1992)
	0	$(5.00 \pm 0.44) \times 10^{-15}$	298	Discharge flow–electron paramagnetic resonance	Orkin and Khamaganov (1993)
	0	3.77×10^{-15}	298	Relative rate (averaged)	DeMore (1993a)
	0	$(3.9 \pm 0.6) \times 10^{-15}$	298	Discharge flow–resonance fluorescence	Leu and Lee (1994)
	0	4.0×10^{-15}	298	Relative rate (averaged)	Barry et al. (1995)
	0	$(4.6 \pm 0.5) \times 10^{-15}$	295	Laser photolysis–ultraviolet absorption	Bednarek et al. (1996)
	0	$(4.9 \pm 1.4) \times 10^{-15}$	294	Discharge flow–resonance fluorescence	Clyne and Holt (1979)
	0	$(2.49 \pm 0.27) \times 10^{-15}$	298	Flash photolysis–resonance absorption	Martin and Paraskevopoulos (1983)
	0	$(2.69 \pm 0.93) \times 10^{-15}$	298	Discharge flow–resonance fluorescence	Brown et al. (1990)
	0	$(1.90 \pm 0.27) \times 10^{-15}$	298	Discharge flow–laser magnetic resonance/pulsed laser photolysis–laser-induced fluorescence	Talukdar et al. (1991)

Table 5 (continued)

Structure	x	Rate constant (cm^3/molecule/s)	T (K)	Method	References
$CF_3(CF_2)_xCHF_2$	0	$(1.64 \pm 0.21) \times 10^{-15}$	298	Relative rate	DeMore (1993b)
	0	$(1.96 \pm 0.25) \times 10^{-15}$	296	Relative rate	Young et al. (2009a)
	2	$(1.85 \pm 0.29) \times 10^{-15}$	296	Relative rate	Young et al. (2009a)
	4	$(1.98 \pm 0.10) \times 10^{-15}$	298	Flash photolysis–laser-induced fluorescence/laser photolysis–laser-induced fluorescence/relative rate	Chen et al. (2004)
	0	$(1.64 \pm 0.28) \times 10^{-15}$	295	Discharge flow–laser-induced fluorescence	Nelson et al. (1993)
	0	$(1.8 \pm 0.2) \times 10^{-15}$	298	Pulsed laser photolysis–laser-induced fluorescence	Zellner et al. (1994)
	0	$(1.62 \pm 0.03) \times 10^{-15}$	298	Flash photolysis–resonance fluorescence	Zhang et al. (1994)
	0	$(1.34 \pm 0.08) \times 10^{-15}$	298	Flash photolysis/pulsed laser photolysis–laser-induced fluorescence	Tokuhashi et al. (2004)
	0	1.80×10^{-15}	298	Relative rate	Hsu and DeMore (1995)

the overall atmospheric fate of CF_3CH_2F (Singh et al. 1996). The kinetics of reaction (31) have been measured in a number of studies using many direct and indirect methods (Table 5) (Barry et al. 1995; Bednarek et al. 1996; Brown et al. 1990; Clyne and Holt 1979; DeMore 1993a; Gierczak et al. 1991; Jeong et al. 1984; Leu and Lee 1994; Liu et al. 1990; Martin and Paraskevopoulos 1983; Orkin and Khamaganov 1993; Zhang et al. 1992). A preferred value has been cited as 4.6×10^{-15} cm^3/molecule/s (Atkinson et al. 2008) from the measurements of Martin and Paraskevopoulos (1983), Liu et al. (1990), Gierczak et al. (1991), Zhang et al. (1992), Orkin and Khamaganov (1993), Leu and Lee (1994), Barry et al. (1995), and Bednarek et al. (1996). Kinetics of reactions are likely to be independent of perfluorinated chain length (Hurley et al. 2004b, c) and apply to all congeners of this class. The tropospheric lifetime with respect to reaction with hydroxyl radicals has been estimated to be between 13.9 and 15.3 years using modeling techniques (Kanakidou et al. 1995; Prather and Spivakovsky 1990; Wild et al. 1996).

The lifetime of transport of CF_3CH_2F to the stratosphere has been estimated as 357 years, suggesting that reactions occurring in the stratosphere will have a minimal impact on the lifetime of this compound (Kanakidou et al. 1995). Studies of the reactivity of CF_3CH_2F in the stratosphere have determined that photolysis and reaction with $O(^1D)$ are insignificant compared to reaction with hydroxyl radicals (Wild et al. 1996). The stratospheric lifetime due to reaction (31) has been determined as 64.8 years (Wild et al. 1996), while the total stratospheric lifetime has been calculated as 62 years (Kanakidou et al. 1995) using two-dimensional and three-dimensional models, respectively.

The overall atmospheric lifetime of CF_3CH_2F has been estimated as 13.8–17.7 years (Houghton et al. 2001; Orlando et al. 1991; Wild et al. 1996). The similarity of this lifetime to that of tropospheric reaction with hydroxyl radicals suggests that this process dominates the atmospheric fate of CF_3CH_2F.

Products of Atmospheric Degradation

Product studies for CF_3CH_2F have focused on atmospheric oxidation by hydroxyl radicals. For these studies, chlorine atoms have typically been used to initiate reactions as a proxy for hydroxyl radicals. Abstraction of a hydrogen atom is followed by reaction with molecular oxygen and NO to yield an alkoxy radical ($CF_3C(O^\bullet)HF$). This radical either reacts with molecular oxygen or decomposes:

$$CF_3C(O^\bullet)HF + O_2 \rightarrow CF_3C(O)F + HO_2 \tag{32a}$$

$$CF_3C(O^\bullet)HF \rightarrow CF_3 + HC(O)F \tag{32b}$$

Study results indicate that both of these processes occur in the troposphere with varying efficacy. Decomposition appears to be dominant in most regions of the atmosphere (Rattigan et al. 1994; Tuazon and Atkinson 1993b; Wallington et al. 1992; Wild et al. 1996). Reaction with molecular oxygen (reaction (32a)) is favored with decreasing temperature and pressure, with the effect of temperature domi-nating (Rattigan et al. 1994). As a result, $CF_3C(O)F$ is produced in higher yield at higher altitudes and latitudes. Near the tropopause, the yield of $CF_3C(O)F$ has been estimated at approximately 0.8 (Rattigan et al. 1994). The area of maximum destruction of CF_3CH_2F is low altitudes of the tropical atmosphere, where approx-imate yields of $CF_3C(O)F$ are lower, between 0.18 and 0.3 (Rattigan et al. 1994; Tuazon and Atkinson 1993b; Wallington et al. 1992; Wild et al. 1996). Total pro-duction of $CF_3C(O)F$ from reaction (32a) has been estimated to range from 0.07 to 0.20 (Wallington et al. 1996). Although absolute yields of $CF_3C(O)F$ and, hence $CF_3C(O)OH$ remain under some debate, it is clear that CF_3CH_2F is a precursor to PFCAs. Longer-perfluorinated-chain congeners of the structure $CF_3(CF_2)_xCH_2F$ may not decompose or react with molecular oxygen in the same ratio. Thus, it is difficult to determine the impact these compounds would have on the PFCA burden.

$CF_3(CF_2)_xCHF_2$

Atmospheric Lifetime

The atmospheric fate of $CF_3(CF_2)_xCHF_2$ has been well studied. The commonly used CF_3CF_2H (HFC-125) has received the most attention in this class. These compounds can react through atmospheric oxidation initiated by chlorine atoms or hydroxyl radicals:

$$CF_3(CF_2)_xCHF_2 + Cl \rightarrow \text{products} \tag{33}$$

$$CF_3(CF_2)_xCHF_2 + OH \rightarrow \text{products} \tag{34}$$

Kinetics of reaction (33) have been measured using relative rate techniques for $CF_3(CF_2)_xCHF_2$, where $x = 0$ (Sehested et al. 1993; Tuazon et al. 1992; Young et al. 2009a) and $x = 2$ (Young et al. 2009a) (Table 5). Values are in reasonable agreement and there is no apparent impact of perfluorinated chain length. A rate constant range can be quoted: k $(Cl + CF_3(CF_2)_xCHF_2) = 2.4 - 2.9 \times 10^{-16}\,cm^2/molecule/s$. Chlorine atoms are not present in the atmosphere to the extent that reaction (33) can impact the atmospheric fate of $CF_3(CF_2)_xCHF_2$ (Singh et al. 1996). Reaction (34) has been studied using both direct measurements and relative rate techniques for $CF_3(CF_2)_xCHF_2$, where $x = 0$ (Brown et al. 1990; Clyne and Holt 1979; DeMore 1992; Martin and Paraskevopoulos 1983; Talukdar et al. 1991; Young et al. 2009a), $x = 2$ (Young et al. 2009a), and $x = 4$ (Chen et al. 2004). There is significant variation among the measurements shown in Table 5. For CF_3CHF_2, a preferred value has been developed that accounts for potential experimental errors of k $(OH + CF_3CHF_2) = 1.9 \times 10^{-15}\,cm^3/molecule/s$ (Atkinson et al. 2008), based on the measurements of Martin and Paraskevopoulos (1983), Talukdar et al. (1991), and DeMore (1992). This preferred value is in agreement with measure-ments made for the longer-chained analogs, suggesting perfluorinated chain length has no impact on kinetics. Atmospheric lifetimes for $CF_3(CF_2)_xCHF_2$ have been determined by scaling the reactivity with hydroxyl radicals to that of methyl chloroform. Brown et al. (1990) determined a lifetime for CF_3CHF_2 of 22.1 years, though this may be underestimated as their measured rate constant is suspected to have been impacted by impurities (Talukdar et al. 1991). Tropospheric lifetimes were determined for CF_3CHF_2 and $CF_3(CF_2)_2CHF_2$ as 33 years (Young et al. 2009a) and for $CF_3(CF_2)_4CHF_2$ as 31 years (Chen et al. 2004). Atmospheric lifetimes for these compounds are expected to be dominated by reaction with hydroxyl radicals, as the compound will not undergo photolysis in the troposphere, and reaction with $O(^1D)$ will be insignificant (Warren et al. 1991). This is supported by agreement of the total atmospheric lifetime of 29 years for CF_3CHF_2 with tropospheric lifetimes determined by hydroxyl radical reaction (Houghton et al. 2001).

Products of Atmospheric Degradation

The atmospheric chemistry of $CF_3(CF_2)_xCHF_2$ has been well studied. Abstraction of the hydrogen atom leads to a perfluorinated radical:

$$CF_3(CF_2)_xCHF_2 + Cl/OH \rightarrow CF_3(CF_2)_x^\bullet + HCL/H_2O \qquad (35)$$

Under most atmospheric conditions, the perfluorinated radical reacts to form $C(O)F_2$, as observed for CF_3CHF_2 (Edney and Driscoll 1992; Olkhov and Smith 2003; Tuazon and Atkinson 1993a). However, under conditions of limited NO_x, the perfluorinated radical can react to form PFCAs as described in *Perfluorinated Radical Mechanism* Section. Smog chamber studies under low-NO_x conditions with offline sample collection demonstrated PFCA formation in low yields from $CF_3(CF_2)_xCHF_2$ ($x = 0,2$) reaction with chlorine atoms. The production of TFA was observed from oxidation of CF_3CHF_2, while formation of perfluorobutanoic acid (PFBA), perfluoropropionic acid (PFPrA), and TFA was observed from reaction of $CF_3CF_2CF_2CHF_2$ (Young et al. 2009a). This demonstrates the potential of $CF_3(CF_2)_xCHF_2$ to act as PFCA precursors under low-NO_x conditions.

$CF_3(CF_2)_xCHFCF_3$

Atmospheric Lifetime

The atmospheric lifetime of CF_3CHFCF_3, known by the trade name HFC-227, has been the subject of a number of studies and is the only compound that has been examined in this class. Reactivity with atmospheric oxidants has been determined:

$$CF_3CHFCF_3 + Cl \rightarrow products \qquad (36)$$

$$CF_3CHFCF_3 + OH \rightarrow products \qquad (37)$$

The kinetics of reaction (36) were determined through direct and indirect methods (Table 5). Measurements are in good agreement, yielding a rate of

k (Cl + CF$_3$CHFCF$_3$) = 3.8–4.5 × 10^{-17} cm^3/molecule/s (Møgelberg et al. 1996; Zellner et al. 1994). Since concentrations of chlorine atoms are low (Singh et al. 1996), reaction (36) is unlikely to impact the atmospheric fate of CF$_3$CHFCF$_3$. Reaction (37) has been measured directly and indirectly (Table 5) (Hsu and DeMore 1995; Nelson Jr. et al. 1993; Tokuhashi et al. 2004; Zellner et al. 1994; Zhang et al. 1994). A recommended value for this reaction is given as k (OH + CF$_3$CHFCF$_3$) = 1.4 × 10^{-15} (cm^3/molecule/s) (Atkinson et al. 2008), based on the values of Zellner et al. (1994), Zhang et al. (1994), and Tokuhashi et al. (2004). Kinetics for longer-chain-length congeners of this class are likely to have similar kinetics, because perfluorinated chain length should not impact the reactivity (Hurley et al. 2004b, c). Reaction with hydroxyl radicals is expected to dominate the atmospheric fate of this compound, and lifetimes with respect to this reaction have been estimated as 34–49 years (Houghton et al. 2001; Møgelberg et al. 1996; Nelson Jr. et al. 1993; Tokuhashi et al. 2004; Zellner et al. 1994; Zhang et al. 1994).

Products of Atmospheric Degradation

The products of reaction (37) for CF$_3$CHFCF$_3$ have been the subject of a single study (Zellner et al. 1994). The product of reaction with hydroxyl radicals is an internal perfluorinated radical that reacts with oxygen and NO to yield an alkoxy radical. The resulting alkoxy radical has two possible decomposition fates, either scission of the C–C or C–F bonds:

$$CF_3C(O^\bullet)FCF_3 \rightarrow CF_3C(O)CF_3 + F \tag{38a}$$

$$CF_3C(O^\bullet)FCF_3 \rightarrow CF_3C(O)F + CF_3 \tag{38b}$$

Studies demonstrated the production of CF$_3$C(O)F in a yield indistinguishable from unity, suggesting that reaction (38b) is the dominant fate of CF$_3$CHFCF$_3$ and reaction (38a) is of minimal significance. As a result, the atmospheric oxidation of CF$_3$CHFCF$_3$ is expected to yield TFA in a yield of 100%.

3.3.2 Hydrofluoroolefins (HFOs)

CF$_3$(CF$_2$)$_x$CF=CH$_2$

Atmospheric Lifetime

The atmospheric fate for a single compound of this class, CF$_3$CF=CH$_2$, has been studied, with respect to tropospheric oxidation:

$$CF_3CF=CH_2 + Cl \rightarrow products \tag{39}$$

$$CF_3CF=CH_2 + OH \rightarrow products \tag{40}$$

$$CF_3CF=CH_2 + O_3 \rightarrow products \tag{41}$$

Kinetics of reaction (39) have been determined using relative rate techniques, yielding a rate of k (Cl + CF$_3$CF=CH$_2$) = (7.03 ± 0.59) × 10^{-11} cm^3/molecule/s (Table 6) (Nielsen et al. 2007). The low abundance of chlorine atoms in the atmosphere (Singh et al. 1996) causes reaction (39) to be of low significance. Reaction (40) has been studied for CF$_3$CF=CH$_2$ by both direct and indirect methods, which are in good agreement (Table 6) (Nielsen et al. 2007; Orkin et al. 1997; Papadimitriou et al. 2008). An approximate rate constant of k (OH + CF$_3$CF= CH$_2$) = 1 × 10^{-12} cm^3/molecule/s can be derived from the measurements. Reactions with ozone may play an important role for unsaturated compounds, for which one measurement has been made, yielding a rate of k (O$_3$ + CF$_3$CF= CH$_2$) = (2.77 ± 0.21) × 10^{-21} cm^3/molecule/s (Table 6) (Nielsen et al. 2007). Using averaged atmospheric abundances, lifetimes with respect to reaction with hydroxyl radical and ozone were estimated as 11 d and 35 years, respectively (Nielsen et al. 2007). Thus, the fate of CF$_3$CF=CH$_2$ is determined by reaction with hydroxyl radical and is on the order of 11 d. The impact of perfluorinated chain length on radical addition reactions in the gas phase is not known, so kinetics of longer congeners could differ. The uncertainty of the lifetimes for CF$_3$CF=CH$_2$ should be stressed, because the high reactivity of these compounds will lead to heterogeneous concentrations throughout the atmosphere and lifetimes that will be defined by local conditions.

Products of Atmospheric Degradation

There have not been any studies that have examined the products of atmospheric oxidation of CF$_3$CF=CH$_2$. However, it has been speculated by Papadimitriou et al. (2008) that a mechanism similar to that described in *Hydrofluoroolefin (HFO) Mechanism* Section occurs for CF$_3$CF=CH$_2$ yielding CF$_3$C(O)F, which would subsequently degrade to form TFA.

CF$_3$(CF$_2$)$_x$CF=CHF

Atmospheric Lifetime

The presence of both Z and E isomers of CF$_3$(CF$_2$)$_x$CF=CHF is expected in the environment, and studies have examined the fate of both isomers for CF$_3$CF=CHF. Atmospheric oxidation of CF$_3$CF=CHF is expected to occur through the following reactions:

$$CF_3CF=CHF + Cl \rightarrow products \qquad (42)$$

$$CF_3CF=CHF + OH + products \qquad (43)$$

$$CF_3CF=CHF + O_3 \rightarrow products \qquad (44)$$

The reaction of chlorine atoms with CF$_3$CF=CHF has been studied using relative rate techniques (Table 6) (Hurley et al. 2007). Kinetics for Z and

Table 6 Summary of kinetics for chlorine atom-, hydroxyl radical- and ozone-initiated reactions of HFOs

	Structure	x	y	Rate constant (cm^3/molecule/s)	T (K)	Method	References
Cl	$CF_3(CF_2)_xCF=CH_2$	0		$(7.03 \pm 0.59) \times 10^{-11}$	296	Relative rate	Nielsen et al. (2007)
	$CF_3(CF_2)_xCF=CHF$	0		$(4.36 \pm 0.48) \times 10^{-11}$ (Z)	296	Relative rate	Hurley et al. (2007)
				$(5.00 \pm 0.56) \times 10^{-11}$ (E)			
	$CF_3(CF_2)_xCF=CF_2$	0		$(2.7 \pm 0.3) \times 10^{-11}$	296	Relative rate	Mashino et al. (2000)
		1		$(1.79 \pm 0.41) \times 10^{-11}$	296	Relative rate	Young et al. (2009b)
	$CF_3(CF_2)_xCF= CF(CF_2)_yCF_3$	0	0	$(7.27 \pm 0.88) \times 10^{-12}$	296	Relative rate	Young et al. (2009b)
	$CF_3(CF_2)_xCF=CH_2$	0		$(1.12 \pm 0.02) \times 10^{-12}$	298	Flash photolysis–resonance fluorescence	Orkin et al. (1997)
		0		$(1.05 \pm 0.17) \times 10^{-12}$	296	Relative rate	Nielsen et al. (2007)
		0		$(1.12 \pm 0.09) \times 10^{-12}$	298	Pulsed laser photolysis–laser-induced fluorescence	Papadimitriou et al. (2008)
	$CF_3(CF_2)_xCF=CHF$	0		$(1.22 \pm 0.14) \times 10^{-12}$ (Z)	296	Relative rate	Hurley et al. (2007)
				$(2.15 \pm 0.23) \times 10^{-12}$ (E)			
		0		$(1.29 \pm 0.06) \times 10^{-12}$ (Z)	298	Pulsed laser photolysis–laser-induced fluorescence	Papadimitriou et al. (2008)

Table 6 (continued)

	Structure	x	y	Rate constant $(cm^3/molecule/s)$	T (K)	Method	References
OH	$CF_3(CF_2)_xCF{=}CF_2$	0		$(2.32 \pm 0.10) \times 10^{-12}$	293	Laser photolysis–laser-induced fluorescence	McIlroy and Tully (1993)
		0		$(2.17 \pm 0.01) \times 10^{-12}$	298	Flash photolysis–resonance fluorescence	Orkin et al. (1997)
		0		$(2.4 \pm 0.3) \times 10^{-12}$	296	Relative rate	Mashino et al. (2000)
		0		$(2.6 \pm 0.7) \times 10^{-12}$	298	Relative rate	Acerboni et al. (2001)
		1		$(1.94 \pm 0.27) \times 10^{-12}$	296	Relative rate	Young et al. (2009b)
	$CF_3(CF_2)_xCF{=}CF(CF_2)_yCF_3$	0	0	$(4.82 \pm 1.15) \times 10^{-13}$	296	Relative rate	Young et al. (2009b)
O_3	$CF_3(CF_2)_xCF{=}CH_2$	0		$(2.77 \pm 0.21) \times 10^{-21}$	296	Relative rate	Nielsen et al. (2007)
	$CF_3(CF_2)_xCF{=}CHF$	0		$(1.45 \pm 0.15) \times 10^{-21}$ (Z) $(1.98 \pm 0.15) \times 10^{-21}$ (E)	296	Relative rate	Hurley et al. (2007)
	$CF_3(CF_2)_xCF{=}CF_2$	0		$(6.2 \pm 1.5) \times 10^{-22}$	298	Pseudo-first-order	Acerboni et al. (2001)

E isomers were indistinguishable within experimental error, with rates of k (Cl + CF$_3$CF = CHF(Z)) = (4.36 ± 0.48) × 10^{-11}cm^3/molecule/s and k (Cl + CF$_3$CF = CHF(E)) = (5.00 ± 0.56) × 10^{-11}cm^3/molecule/s. Of more atmospheric relevance is reaction (43), which was examined using relative rate methods for both isomers and using pulsed laser photolysis–laser-induced fluorescence for the Z isomer (Table 6) (Hurley et al. 2007; Papadimitriou et al. 2008). Results for the Z isomer are in good agreement, yielding an approximate rate of k (OH + CF$_3$CF = CHF(Z)) = 1.22 − 1.29 × 10^{-12} cm^3/molecule/s, while the rate for the E isomer is significantly higher: k (OH + CF$_3$CF=CHF(E)) = (2.15 ± 0.23) × 10^{-12} cm^3/molecule/s. By scaling the rate of reaction to that of CH$_3$CCl$_3$, approximated lifetimes of 18 and 10 d for the Z and E isomers, respectively, were determined (Hurley et al. 2007). A similarly short lifetime of 10 d was estimated for reaction of the Z isomer with hydroxyl radicals using a globally averaged concentration of hydroxyl radicals (Papadimitriou et al. 2008). The lifetimes for this compound are approximate, due to the short lifetime and dependence on local conditions. The rate of reaction with ozone has also been determined, for which values again differed between the two isomers, with rates of k (O$_3$ + CF=CHF(Z)) = (1.45 ± 0.15) × 10^{-21} cm^3/molecule/s and k (O$_3$ + CF$_3$CF = CHF(E)) = (1.98 ± 0.15) × 10^{-21} cm^3/molecule/s (Hurley et al. 2007). Using an approximate background ozone concentration of 35 ppb, lifetimes for Z and E isomers were estimated as 25 and 1.9 years (Hurley et al. 2007). Thus, the atmospheric lifetime of CF$_3$CF=CHF will be dominated by reaction with hydroxyl radicals and will be on the order of 10–20 d.

Products of Atmospheric Degradation

Studies of the products of reactions (42) and (43) have been performed on both Z and E isomers of CF$_3$CF=CHF (Hurley et al. 2007). Reactions initiated with chlorine atoms or hydroxyl radicals in the presence or absence of NO$_x$ all yielded CF$_3$C(O)F and HC(O)F in molar yields. This occurs by the mechanism described in *Hydrofluoroolefin (HFO) Mechanism* Section, indicating that CF$_3$(CF$_2$)$_x$CF=CHF forms perfluoroacyl fluorides, and subsequently, PFCAs in a yield of unity.

CF$_3$(CF$_2$)$_x$CF=CF$_2$

Atmospheric Lifetime

Studies have examined the fate of CF$_3$(CF$_2$)$_x$CF=CF$_2$ with atmospheric oxidants:

$$CF_3(CF_2)_xCF=CF_2 + Cl \rightarrow products \tag{45}$$

$$CF_3(CF_2)_xCF=CF_2 + OH \rightarrow products \tag{46}$$

$$CF_3(CF_2)_xCF=CF_2 + O_3 \rightarrow products \tag{47}$$

Kinetics for reaction (45) have been studied for $CF_3(CF_2)_xCF{=}CF_2$ for $x = 0$ (Mashino et al. 2000) and $x = 1$ (Young et al. 2009b), using relative rate methods (Table 6). Significant differences exist between the two measurements, suggesting an impact of perfluorinated chain length on the reaction kinetics. The low concentrations of chlorine atoms in the atmosphere (Singh et al. 1996) indicate that reaction (45) is not a significant contributor to the atmospheric fate of $CF_3(CF_2)_xCF{=}CF_2$. Reaction of $CF_3(CF_2)_xCF{=}CF_2$ with hydroxyl radicals has been studied using direct and indirect methods for $CF_3(CF_2)_xCF{=}CF_2$, $x = 0$ (Acerboni et al. 2001; Mashino et al. 2000; McIlroy and Tully 1993; Orkin et al. 1997) and $x = 1$ (Young et al. 2009b) (Table 6). Reported rate constants for both compounds were within reasonable agreement, though it appears as though the longer-perfluorinated chain may result in slightly lower reactivity. Rate constants were determined as $k\,(OH{+}CF_3CF{=}CF_2) = 2.17{-}2.6 \times 10^{-12}$ cm^3/molecule/s and $k\,(OH{+}CF_3CF_2CF{=}CF_2) = (1.94 \pm 0.27) \times 10^{-12}$ cm^3/molecule/s. Lifetimes with respect to reaction (46) determined either using a globally averaged concentration of hydroxyl radicals or by comparison to CH_3CCl_3 days were 6–9 d for $CF_3CF{=}CF_2$ (Acerboni et al. 2001; Mashino et al. 2000) and 6 d for $CF_3CF_2CF{=}CF_2$ (Young et al. 2009b). A single measurement for reaction (47) with $CF_3CF{=}CF_2$ has been made using pseudo-first-order techniques (Acerboni et al. 2001). A rate of $k\,(O_3 + CF_3CF{=}CF_2) = (6.2 \pm 1.5) \times 10^{-22}$ cm^3/molecule/s yields a lifetime on the order of several years. The atmospheric fate of $CF_3(CF_2)_xCF{=}CF_2$ is dominated by reaction with hydroxyl radicals, with lifetimes of about a week.

Products of Atmospheric Degradation

Products of reactions with chlorine atoms and hydroxyl radicals have been studied for $CF_3(CF_2)_xCF{=}CF_2$, where $x = 0$ (Acerboni et al. 2001; Mashino et al. 2000) and $x = 1$ (Young et al. 2009b). Reactions of $CF_3CF{=}CF_2$ in the presence of NO_x gave $CF_3C(O)F$ and $C(O)F_2$ in approximately 100% yield when initiated by chlorine atoms or hydroxyl radicals. Similarly, $CF_3CF_2C(O)F$ and $C(O)F_2$ were each observed in a yield of unity from the atmospheric oxidation of $CF_3CF_2CF{=}CF_2$ in the presence and absence of NO_x. These products are formed through the reactions described in *Hydrofluoroolefin (HFO) Mechanism* Section and confirm $CF_3(CF_2)_xCF{=}CF_2$ as precursors of PFCAs.

$CF_3(CF_2)_xCF{=}CF(CF_2)_xCF_3$

Atmospheric Lifetime

The atmospheric fate of one compound from this class, $CF_3CF{=}CFCF_3$, has been examined with respect to reaction with chlorine atoms and hydroxyl radicals:

$$CF_3CF{=}CFCF_3 + Cl \rightarrow products \qquad (48)$$

$$CF_3CF{=}CFCF_3 + OH \rightarrow products \qquad (49)$$

Kinetics for reactions (48) and (49) were determined using relative rate techniques, yielding rate constants of k (Cl + CF$_3$CF=CFCF$_3$) = (7.27 ± 0.88) × 10^{-12} cm^3/molecule/s and k (OH + CF$_3$CF = CFCF$_3$) = (4.82 ± 1.15) × 10^{-13} cm^3/molecule/s (Young et al. 2009b). There have been no experimental measurements of reaction with ozone, but by proxy to other HFOs, this reaction is expected to be of little to no significance to the fate of these compounds. Hence, the atmospheric lifetime of CF$_3$CF=CFCF$_3$ is determined by reaction with hydroxyl radicals and is estimated to be 24 d, using a globally averaged hydroxyl radical concentration (Young et al. 2009b).

Products of Atmospheric Degradation

The products of reaction of CF$_3$CF=CFCF$_3$ with chlorine atoms and hydroxyl radicals have been examined by Young et al. (2009b) in both the presence and absence of NO$_x$. In all cases, CF$_3$C(O)F was observed as the sole product, in a yield of approximately 200%. This occurs through the mechanism described in *Hydrofluoroolefin (HFO) Mechanism* Section and confirms this compound class as PFCA precursors.

3.4 Fluorotelomer and Related Compounds

Numerous fluorotelomer compounds have been produced commercially, including fluorotelomer iodides (FTIs), fluorotelomer olefins (FTOs), fluorotelomer alcohols (FTOHs), and fluorotelomer acrylates (FTAcs). These compounds are named with the prefix *x:y*, where *x* is the number of perfluorinated carbons and *y* is the number of hydrogenated carbons. For example, 6:2 FTOH refers to C$_6$F$_{13}$CH$_2$CH$_2$OH. Among the compounds that are confirmed to be high-production volume chemicals are 4:2–18:2 FTIs, 4:2 FTO, 2:2–18:2 FTOHs, and 6:2–14:2 FTAcs (Howard and Meylan 2007). Other compounds described below, including perfluorinated aldehydes and aldehyde hydrates and fluorotelomer aldehydes, are known products of the degradation of these commercially produced compounds and have themselves been intensively studied.

3.4.1 Perfluorinated Aldehyde (PFAL) Hydrates

Atmospheric Lifetime

The PFAL hydrates (CF$_3$(CF$_2$)$_x$C(OH)$_2$) are a stable form of PFALs and are the form in which PFALs can be commercially purchased. The stability of these hydrates is demonstrated by the rigorous dehydration required to yield the PFAL, consisting of distillation in the presence of phosphate pentoxide (Sulbaek Andersen et al. 2006). The increased polarity of the PFAL hydrate relative to PFAL suggests that rain-out of the atmosphere could occur with these compounds. However, nuclear magnetic resonance (NMR) studies have demonstrated the loss of $CF_3CF_2C(OH)_2$ from an aqueous-phase sample, suggesting a moderate air–water partitioning coefficient for the hydrate (Sulbaek Andersen et al. 2006). With increasing chain length, air–water partitioning would be expected to increase. Thus, gas-phase reactivity of PFAL hydrates is of atmospheric relevance. The reactivity of PFAL hydrates has been examined in a single study (Sulbaek Andersen et al. 2006). These compounds may be subject to atmospheric oxidation with chlorine radicals or hydroxyl radicals:

$$CF_3(CF_2)_xC(OH)_2 + Cl \rightarrow \text{products} \tag{50}$$

$$CF_3(CF_2)_xC(OH)_2 + OH \rightarrow \text{products} \tag{51}$$

The kinetics of reaction (50) were studied using relative rate techniques at 296 K for $CF_3(CF_2)_xC(OH)_2$ ($x = 0, 2, 3$) and were determined to be independent of chain length for the molecules studied. A final rate constant was cited as $k(Cl + CF_3(CF_2)_xC(OH)_2) = (5.84 \pm 0.92) \times 10^{-13}$ cm^3/molecule/s (Table 7). Similarly, a rate constant was determined for reaction (51) for $CF_3C(OH)_2$, with a quoted rate constant of $(1.22 \pm 0.26) \times 10^{-13}$ cm^3/molecule/s (Table 8).

PFAL hydrates are expected to exist in equilibrium with PFALs in the gas phase:

$$CF_3(CF_2)_xC(OH)_2 \rightleftharpoons CF_3(CF_2)_xC(O)H + H_2O \tag{52}$$

Equilibrium (52) was investigated in a dry smog chamber environment, in which gas-phase $CF_3C(OH)_2$ was observed to lose water slowly, forming $CF_3C(O)H$ with a yield close to unity (Sulbaek Andersen et al. 2006). The authors speculate that dehydration may be a result of heterogeneous processes, in which case smog chamber conditions are difficult to directly relate to the atmosphere.

Typical chlorine atom concentrations are too low to affect the atmospheric lifetimes of organic contaminants (Singh et al. 1996). Using a globally averaged hydroxyl radical concentration of 1×10^6 molecules/cm^3, an approximate atmospheric lifetime with respect to hydroxyl radical reaction of 90 d was determined for PFAL hydrates (Sulbaek Andersen et al. 2006). It is possible that conversion to FTALs or partitioning could occur on a faster timescale than reaction with hydroxyl radicals. Due to uncertainties regarding these processes, it is not currently possible to determine an overall atmospheric lifetime for PFAL hydrates.

Products of Atmospheric Degradation

Products of reaction (50) were studied in a smog chamber setting for $CF_3C(OH)_2$ and TFA was observed to be the sole primary product (Sulbaek Andersen et al. 2006). The production of TFA from reaction (51) was also observed using smog chamber techniques, though the yield was not quantified. The mechanism by which PFCAs are hypothesized to form from PFAL hydrates is discussed in *Perfluorinated Aldehyde (PFAL) Hydrate Mechanism* Section.

Table 7 Summary of chlorine atom–initiated kinetics for fluorotelomer and related compounds

Structure	x	Rate constant (cm^3/molecule/s)	T (K)	Method	References
PFAL hydrate ($CF_3(CF_2)_xCH_2C(OH)_2$)	0,2,3	$(5.84 \pm 0.92) \times 10^{-13}$	296	Relative rate	Sulbaek Andersen et al. (2006)
PFAL ($CF_3(CF_2)_xC(O)H$)	0	2.28×10^{-12}	298	Relative rate	Scollard et al. (1993)
		$(1.8 \pm 0.4) \times 10^{-12}$	295	Relative rate	Wallington and Hurley (1993)
		$(1.90 \pm 0.25) \times 10^{-12}$	296	Fitting	Hurley et al. (2004b)
		$(1.85 \pm 0.26) \times 10^{-12}$	296	Relative rate	Sulbaek Andersen et al. (2004a)
	1	$(1.96 \pm 0.28) \times 10^{-12}$	296	Relative rate	Sulbaek Andersen et al. (2003b)
		$(2.35 \pm 0.42) \times 10^{-12}$	296	Fitting	Hurley et al. (2004b)
	2	$(2.56 \pm 0.35) \times 10^{-12}$	296	Fitting	Hurley et al. (2004b)
		$(2.03 \pm 0.23) \times 10^{-12}$	296	Relative rate	Sulbaek Andersen et al. (2004a)
	3	$(2.48 \pm 0.31) \times 10^{-12}$	296	Fitting	Hurley et al. (2004b)
		$(2.34 \pm 0.25) \times 10^{-12}$	296	Relative rate	Sulbaek Andersen et al. (2004a)
	5	$(2.8 \pm 0.7) \times 10^{-12}$	298	Relative rate	Solignac et al. (2006a)
FTAL ($CF_3(CF_2)_xCH_2C(O)H$)	0	$(2.57 \pm 0.04) \times 10^{-11}$	298	Relative rate	Kelly et al. (2005)
		$(1.81 \pm 0.27) \times 10^{-11}$	296	Relative rate	Hurley et al. (2005)
	3	$(1.84 \pm 0.30) \times 10^{-11}$	296	Fitting	Hurley et al. (2004a)
	7	$(1.9 \pm 0.2) \times 10^{-11}$	296	Relative rate	Chiappero et al. (2008)
oFTOH ($CF_3(CF_2)_xCH_2OH$)	0,2,3	$(6.48 \pm 0.53) \times 10^{-13}$	296	Relative rate	Hurley et al. (2004b)
	0	$(6.78 \pm 0.63) \times 10^{-13}$	303	Very low-pressure reactor mass spectrometry	Papadimitriou et al. (2007)

Table 7 (continued)

Structure	x	Rate constant (cm^3/molecule/s)	T (K)	Method	References
FTOH ($CF_3(CF_2)_xCH_2CH_2OH$)	0	$(2.24 \pm 0.04) \times 10^{-11}$	296	Relative rate	Kelly et al. (2005)
		$(1.59 \pm 0.20) \times 10^{-11}$	296	Relative rate	Hurley et al. (2005)
		$(1.90 \pm 0.17) \times 10^{-11}$	303	Very low-pressure reactor mass spectrometry	Papadimitriou et al. (2007)
	3,5,7	$(1.61 \pm 0.49) \times 10^{-11}$	296	Relative rate	Ellis et al. (2003)
FTO	0,1,3,5,7	$(9.07 \pm 1.08) \times 10^{-11}$	296	Relative rate	Sulbaek Andersen et al. (2005b)
($CF_3(CF_2)_xCH=CH_2$)					
FTI ($CF_3(CF_2)_xCH_2CH_2I$)	3	$(1.25 \pm 0.15) \times 10^{-12}$	295	Relative rate	Young et al. (2008)
FTAc ($CF_3(CF_2)_xCH_2CH_2$ OC(O)CHCH_2$)	3	$(2.21 \pm 0.16) \times 10^{-10}$	296	Relative rate	Butt et al. (2009)

Table 8 Summary of hydroxyl radical-initiated kinetics for fluorotelomer and related compounds at atmospheric pressure

Structure	x	Rate constant (cm^3/molecule/s)	T (K)	Method	References
PFAL hydrate $(CF_3(CF_2)_xCH_2C(OH)_2)$	0	$(1.22 \pm 0.26) \times 10^{-13}$	296	Relative rate	Sulbaek Andersen et al. (2006)
PFAL $(CF_3(CF_2)_xC(O)H)$	0	$(1.1 \pm 0.7) \times 10^{-12}$	299	Discharge flow–resonance fluorescence	Dobe et al. (1989)
		$(6.5 \pm 0.5) \times 10^{-13}$	298	Pulsed laser photolysis–resonance fluorescence	Scollard et al. (1993)
		$(5.4 \pm 1.2) \times 10^{-13}$	298	Relative rate	Scollard et al. (1993)
		$(4.80 \pm 0.31) \times 10^{-13}$	298	Relative rate	Sellevåg et al. (2004)
	1	$(5.26 \pm 0.80) \times 10^{-13}$	296	Relative rate	Sulbaek Andersen et al. (2003b)
	0,2,3	$(6.5 \pm 1.2) \times 10^{-13}$	296	Relative rate	Sulbaek Andersen et al. (2004a)
	2	$(5.8 \pm 0.6) \times 10^{-13}$	298	Pulsed laser photolysis–laser-induced fluorescence	Solignac et al. (2007)
	3	$(6.1 \pm 0.5) \times 10^{-13}$	298	Pulsed laser photolysis–laser-induced fluorescence	Solignac et al. (2007)
FTAL $(CF_3(CF_2)_xCH_2C(O)H)$	0	$(3.30 \pm 0.08) \times 10^{-12}$	298	Relative rate	Sellevåg et al. (2004)
		$(2.96 \pm 0.04) \times 10^{-12}$	298	Pulsed laser photolysis–laser-induced fluorescence	Kelly et al. (2005)
		$(2.57 \pm 0.44) \times 10^{-12}$	296	Relative rate	Hurley et al. (2005)
	7	$(2.0 \pm 0.4) \times 10^{-12}$	296	Relative rate	Chiappero et al. (2008)

Table 8 (continued)

Structure	x	Rate constant (cm^3/molecule/s)	T (K)	Method	References
oFTOH ($CF_3(CF_2)_xCH_2OH$)	1	1.15×10^{-13}	298	Reactor-GC–MS	Chen et al. (2000)
	0,2,3	$(1.02 \pm 0.10) \times 10^{-13}$	296	Relative rate	Hurley et al. (2004b)
	0	$(1.08 \pm 0.05) \times 10^{-12}$	298	Relative rate	Kelly et al. (2005)
FTOH ($CF_3(CF_2)_xCH_2CH_2OH$)		$(8.9 \pm 0.3) \times 10^{-13}$	298	Pulsed laser photolysis–laser-induced fluorescence	Kelly et al. (2005)
	3,5,7	$(6.91 \pm 0.91) \times 10^{-13}$	296	Relative rate	Hurley et al. (2005)
		$(1.07 \pm 0.22) \times 10^{-12}$	296	Relative rate	Ellis et al. (2003)
	5	$(7.9 \pm 0.8) \times 10^{-13}$	298	Relative rate	Kelly et al. (2005)
FTO ($CF_3(CF_2)_xCH{=}CH_2$)	0	$(1.54 \pm 0.05) \times 10^{-12}$	298	Flash photolysis–resonance fluorescence	Orkin et al. (1997)
	5	$(1.35 \pm 0.11) \times 10^{-12}$	298	Pulsed laser photolysis–laser-induced fluorescence	Vesine et al. (2000)
FTI ($CF_3(CF_2)_xCH_2CH_2I$)	0,1,3,5,7	$(1.36 \pm 0.25) \times 10^{-12}$	296	Relative rate	Sulbaek Andersen et al. (2005b)
FTAc ($CF_3(CF_2)_x$ $CH_2CH_2OC(O)CHCH_2$)	3	$(1.2 \pm 0.6) \times 10^{-12}$	295	Relative rate	Young et al. (2008)
	3	$(1.13 \pm 0.12) \times 10^{-11}$	296	Relative rate	Butt et al. (2009)

3.4.2 Perfluorinated Aldehydes (PFALs)

Atmospheric Lifetime

The atmospheric fate of PFALs involves contributions from a number of processes. The first is reaction with chlorine atoms or hydroxyl radicals:

$$CF_3(CF_2)_xC(O)H + Cl \rightarrow products \tag{53}$$

$$CF_3(CF_2)_xC(O)H + OH \rightarrow products \tag{54}$$

The kinetics of reaction (53) have been studied using relative rate methods (Table 7) (Scollard et al. 1993; Solignac et al. 2006b; Sulbaek Andersen et al. 2003b, 2004a; Wallington and Hurley 1993) and fitting techniques (Hurley et al. 2004b). A slight increase in rate constant was observed with increasing perfluorinated chain length, although this did not exceed the error associated with the individual determinations. The reaction of PFALs of all chain lengths with chlorine atoms proceeds with rate constant, $k(Cl + CF_3(CF_2)C(O)H)$, of $(1.8 - 2.8) \times 10^{-12}$ cm^3/molecule/s. The kinetics of reaction (54) for $CF_3C(O)H$ have been studied in detail using a number of techniques (Table 8) (Dobe et al. 1989; Scollard et al. 1993; Sellevåg et al. 2004; Sulbaek Andersen et al. 2004a). An IUPAC preferred value is given as $k(OH + CF_3C(O)H) = 5.7 \times 10^{-13}$ cm^3/molecule/s at 298 K (Atkinson et al. 2008), which is the average of values reported by Scollard et al. (1993), Sellevåg et al. (2004), and Sulbaek Andersen et al. (2004a). Longer-chain PFALs have been studied using relative rate techniques for $CF_3(CF_2)_xC(O)H$ ($x = 0 - 3$) (Sulbaek Andersen et al. 2003b, 2004a) (Table 8) and laser photolysis–laser-induced fluorescence for $CF_3(CF_2)_xC(O)H$ ($x = 2, 3$) (Solignac et al. 2007). Apart from one measurement for $CF_3C(O)H$ (Dobe et al. 1989) that was excluded from the IUPAC-recommended value due to large uncertainty (Atkinson et al. 2008), rate constants for PFALs of different chain lengths are indistinguishable from one another, within error. Thus, the rate of reaction of PFALs with hydroxyl radicals is $k(OH + CF_3(CF_2)_xC(O)H) = (0.58 - 1.1) \times 10^{-12}$ cm^3/molecule/s.

A further atmospheric sink of PFALs is reaction with water to form stable hydrates:

$$CF_3(CF_2)_xC(O)H + H_2O \rightleftharpoons CF_3(CF_2)_xC(OH)_2 \tag{55}$$

An investigation into the likelihood of hydration of PFALs was undertaken by Sulbaek Andersen et al. (2006). Observations indicated that reaction (55) did not occur under homogeneous gas-phase conditions, with an upper limit for reaction calculated as $k(CF_3C(O)H_{(g)} + H_2O_{(g)}) < 2 \times 10^{-23}$ cm^3/molecule/s. However, it was demonstrated that gas-phase PFALs were lost rapidly upon contact with liquid water and that some of the PFALs formed PFAL hydrates. In humid smog chamber experiments, gas-phase $CF_3C(O)H$ slowly formed $CF_3C(OH)_2$. Rates between replicate experiments were inconsistent, which suggest that the mechanism of hydration is heterogeneous. Because information on surfaces within the smog chamber is not available, these experiments cannot be quantitatively applied to the atmosphere, though they do shed some light on the nature of equilibrium between PFALs and PFAL hydrates.

Finally, PFALs can also be subject to photolysis:

$$CF_3(CF_2)_xC(O)H + h\nu \rightarrow \text{products} \qquad (56)$$

A few studies have examined the potential of photolysis to limit the atmospheric lifetime of PFALs (Chiappero et al. 2006; Sellevåg et al. 2004; Solignac et al. 2007). Observations regarding the UV absorption cross-sections for $CF_3(CF_2)_xC(O)H$ ($x = 0 - 3$) are in good agreement (Table 9). The measurements suggest a red-shift in absorption for PFALs in moving from $CF_3C(O)H$ to $CF_3CF_2C(O)H$, but no effect in further increasing perfluorinated chain length. There is also a clear increase in UV absorption cross-section with increasing perfluorinated chain length. Chiappero et al. (2006) determined quantum yields of dissociation for PFALs of four perfluorinated chain lengths. A decrease in quantum yield of dissociation was observed with increasing perfluorinated carbons, presumably as a result of the greater degrees of freedom in the larger molecules. However, measurements of photolysis quantum yields and ultimate photolysis lifetimes are not consistent between studies. Chiappero et al. (2006) determined photolytic lifetimes by interpolating quantum yields over the absorbance range. This, along with measured UV absorption cross-sections, was entered into the Tropospheric Ultraviolet–Visible (TUV) model to yield approximate lifetimes of <2 d for $CF_3(CF_2)_xC(O)H$ ($x = 1, 2, 3$) and <6 d for $x = 0$. Similar lifetimes were determined by Solignac et al. (2007), with lifetimes on the order of hours to days measured for $CF_3(CF_2)_xC(O)H$ ($x = 2, 3, 5$) in the Euphore chamber in Valencia, Spain under natural sunlight conditions. There is significant discrepancy between these results and those of Sellevåg et al. (2004), who determined an atmospheric lifetime for $CF_3C(O)H$ of >27 d, also using the Euphore chamber. The source of this inconsistency is not clear. Although PFALs are much more reactive with chlorine atoms than with hydroxyl radicals, typical concentrations of chlorine are too low to affect overall atmospheric fate (Singh et al. 1996). Assuming an average concentration of hydroxyl radicals of 1×10^6 molecules/cm^3, the lifetime of PFALs with respect to the reaction with atmospheric oxidants ranges from about 10 to 20 d. Of the processes that are currently understood for PFALs, photolysis dominates with the majority of studies suggesting that degradation occurs over the timescale of less than 2 d. More studies are required to determine the relative importance of photolysis and formation of PFAL hydrates in the overall atmospheric fate of PFALs.

Products of Atmospheric Degradation

The products of atmospheric degradation have been well studied for PFALs (Chiappero et al. 2006; Hurley et al. 2006; Solignac et al. 2006b; Sulbaek Andersen et al. 2003a, b, 2004a, b). Reaction of PFALs with chlorine atoms and hydroxyl radicals is known to yield perfluoroacyl radicals:

Table 9 Photolysis properties of perfluorinated aldehydes (PFALs) and fluorotelomer aldehydes (FTALs)

x	Absorption max (nm)	Cross-section at absorption max ($\times 10^{-20}$ cm^2/molecule)	Quantum yield of dissociation (measured wavelength [nm])	Atmospheric lifetime	References
PFAL (CF$_3$(CF$_2$)$_x$C(O)H)					
0	300	2.89	0.79 (254); 0.17 (308)	<6 d[a]	Chiappero et al. (2006)
	301	3.2	<0.02 (full range of atmospheric absorption)	>27 d[b]	Sellevåg et al. (2004)
1	308	5.86	0.81 (254)	<2 d[a]	Chiappero et al. (2006)
2	308	8.15	0.63 (254)	<2 d[a]	Chiappero et al. (2006)
	309	8.1 ± 0.6	0.023 (range of atmospheric absorption)	21 ± 10 h[b]	Solignac et al. (2007)
3	308	9.49	0.60 (254); 0.08 (308)	<2 d[a]	Chiappero et al. (2006)
	309	9.4 ± 0.7	0.029 (full range of atmospheric absorption)	15 ± 7 h[b]	Solignac et al. (2007)
5	290	3.52	0.74 (254); 0.04 (308)	46 ± 23 h[b]	Solignac et al. (2007)
FTAL (CF$_3$(CF$_2$)$_x$CH$_2$C(O)H)					
0	292	3.845	<0.04 (full range of atmospheric absorption)	<40 d[a]	Chiappero et al. (2006)
				>15 d[b]	Sellevåg et al. (2004)
5	300	13.3	0.55 (254)	<20 d[a]	Chiappero et al. (2006)
	283	5.4 ± 0.4			Solignac et al. (2007)

[a]Determined using TUV model
[b]Measured in Euphore chamber

$$CF_3(CF_2)_xC(O)H + Cl/OH \rightarrow CF_3(CF_2)_xC(O)^\bullet + HCl/H_2O \qquad (57)$$

Two major fate pathways exist for the perfluoroacyl radical: reaction with oxygen and loss of carbon monoxide to yield a perfluorinated radical:

$$CF_3(CF_2)_xC(O)^\bullet + O_2 \rightarrow CF_3(CF_2)_xC(O)OO^\bullet \qquad (58a)$$

$$CF_3(CF_2)_xC(O)^\bullet + M \rightarrow CF_3(CF_2)_x^\bullet + CO + M \qquad (58b)$$

Reaction (58a) has been demonstrated to occur through the observation of PFCA formation from PFALs (see *Perfluoroacyl Peroxy Radical Mechanism* Section) (Ellis et al. 2004; Hurley et al. 2004a, 2006; Sulbaek Andersen et al. 2004b). The prevalence of reactions (58a) and (58b) has been studied both experimentally (Hurley et al. 2006; Solignac et al. 2006b) and computationally (Mereau et al. 2001; Waterland and Dobbs 2007). As expected, experimental yields of reaction (58a) increase with increasing concentration of oxygen (Hurley et al. 2006). Product yields of reaction (58b) (Table 10) in air were shown to increase with increasing perfluorinated chain length (Hurley et al. 2006). Computational studies demonstrated that increasing perfluorinated chain length weakens the C–CO bond, leading to an increased prevalence of the decomposition reaction (58b) (Waterland and Dobbs 2007). Regardless of the mechanism, chlorine atom- or hydroxyl radical-initiated reaction of PFALs can lead to the formation of PFCAs, whether through reaction (58a) and the perfluoroacyl peroxy mechanism (*Perfluoroacyl Peroxy Radical Mechanism* Section) or through reaction (58b) and the perfluorinated radical mechanism (*Perfluorinated Radical Mechanism* Section).

Products of photolysis of PFALs were studied by Chiappero et al. (2006). Two products were observed from photolysis reactions of PFALs $(CF_3(CF_2)_xC(O)H, x = 0 - 3)$ at 254 nm: $CF_3(CF_2)_x^\bullet$ radicals and $CF_3(CF_2)_xH$, suggesting occurrence of the following reactions:

$$CF_3(CF_2)_xC(O)H + h\nu \rightarrow CF_3(CF_2)_x^\bullet + HCO \qquad (59a)$$

$$CF_3(CF_2)_xC(O)H + h\nu \rightarrow CF_3(CF_2)_xH + CO \qquad (59b)$$

Table 10 Fate of perfluoroacyl radicals at atmospheric oxygen levels

R	Product yields		References
	$RC(O) \rightarrow R + CO$	$RC(O) + O_2 \rightarrow RC(O)OO$	
CF_3	0.02	0.98	Mereau et al. (2001)
	0.02	0.98	Hurley et al. (2006)
CF_3CF_2	0.52	0.48	Hurley et al. (2006)
	0.61	0.39	Solignac et al. (2006a)
$CF_3CF_2CF_2$	0.81	0.19	Hurley et al. (2006)
$CF_3CF_2CF_2CF_2$	0.89	0.11	Hurley et al. (2006)

Both reactions appeared to occur with approximately equal prevalence at 254 nm. However, actinic radiation is not available at 254 nm in the lower atmosphere. Experiments at 308 nm, which is close to the absorption maximum for PFALs and within the tropospheric actinic spectrum, did not observe the formation of $CF_3(CF_2)_xH$ from photolysis of PFALs ($CF_3(CF_2)_xC(O)H$, $x = 0 - 3$). Irradiation experiments of $CF_3C(O)H$ at 308 nm used nitric oxide (NO) as a radical scavenger for perfluorinated radicals. Under these conditions, a $98 \pm 7\%$ yield for CF_3NO resulted from the reaction of CF_3 with NO, suggesting that reaction (59a) is prevalent at this wavelength.

The dominant product of PFAL degradation is likely the photolysis product. Photolysis of PFALs yields primarily the corresponding perfluorinated radical, which has been shown to form PFCAs of all chain lengths (*Perfluorinated Radical Mechanism* Section).

3.4.3 Fluorotelomer Aldehydes (FTALs)

Atmospheric Lifetime

The atmospheric lifetime of FTALs can be limited by reaction with chlorine atoms and hydroxyl radicals:

$$CF_3(CF_2)_xCH_2C(O)H + Cl \rightarrow products \qquad (60)$$

$$CF_3(CF_2)_xCH_2C(O) + OH \rightarrow products \qquad (61)$$

Reaction (60) has been measured by relative rate techniques (Chiappero et al. 2008; Hurley et al. 2005; Kelly et al. 2005) and determined through fitting product formation and degradation curves (Hurley et al. 2004a) (Table 7). There is good agreement between the values, with the exception of that measured by Kelly et al. (2005). Hurley et al. (2005) suggest that this discrepancy may be due to the use of a single reference compound and errors associated with the reference rate. In addition, there does not appear to be a chain-length effect on the rate of the reaction. A final rate constant can be determined from the values in Table 7, excluding the outlying value of Kelly et al. (2005) and including extremes of the individual measurements of $k(Cl + CF_3(CF_2)_xCH_2C(O)H) = (1.85 \pm 0.31) \times 10^{-11} cm^3/molecule/s$. The kinetics of reaction (61) have been studied by relative rate techniques (Chiappero et al. 2008; Hurley et al. 2005; Sellevåg et al. 2004) and pulsed laser photolysis–laser-induced fluorescence techniques (Kelly et al. 2005) (Table 8). Within the three measurements of $CF_3CH_2C(O)H$, there is significant discrepancy, the source of which is unclear. The quoted rate for longer-perfluorinated-chain FTALs is somewhat lower, though within error of some of the rates for $CF_3CH_2C(O)H$, suggesting that, as with other polyfluorinated compounds, perfluorinated chain length does not impact kinetics.

The photolysis of FTALs may also contribute to their atmospheric fate:

$$CF_3(CF_2)_xCH_2C(O)H + h\nu \rightarrow products \tag{62}$$

The potential for FTALs to undergo photolysis has been examined in three studies (Chiappero et al. 2006; Sellevåg et al. 2004; Solignac et al. 2007) (Table 9). The researchers who studied the fate of $CF_3CH_2C(O)H$ are in agreement, both in regards to the absorption cross-section and the quantum yield of dissociation. Chiappero et al. (2006) used the TUV model to determine a lifetime for $CF_3CH_2C(O)H$ of less than 40 d, while Sellevåg et al. (2004) measured a lifetime of greater than 15 d in the Euphore chamber. Although the photolysis of $CF_3(CF_2)_5CH_2C(O)H$ has also been examined, the measurements show poor agreement. It can be seen from Table 9 that the absorption cross-section and location of maximum absorption vary significantly between the two studies. Chiappero et al. (2006) determined a quantum yield of dissociation, but at a wavelength well into the UV and below the absorption maximum for the chromophore. Nevertheless, they used the TUV model to estimate a lifetime with respect to photolysis of less than 20 d.

Atmospheric concentrations of chlorine atoms are too small to impact the atmospheric fate of FTALs (Singh et al. 1996). Assuming an average concentration of hydroxyl radicals of 9.4×10^5 molecules/cm^3, Sellevåg et al. (2004) determined a lifetime of approximately 4 d for reaction of $CF_3CH_2C(O)H$. Although there are discrepancies within the limited available data concerning the atmospheric fate of FTALs, it appears as though the atmospheric lifetime will be determined by reaction with hydroxyl radical and will be on the order of a few days.

Products of Atmospheric Degradation

There have been few studies in which the products of atmospheric reaction of FTALs have been examined. The products of reaction (60) have been studied for $CF_3CH_2C(O)H$ and $CF_3(CF_2)_3CH_2C(O)H$ under smog chamber conditions in the absence of NO_x (Hurley et al. 2004a, 2005). A dominant primary product for both FTALs was the corresponding PFAL, with a yield of $46 \pm 3\%$ in the case of $CF_3(CF_2)_3C(O)H$ formed from $CF_3(CF_2)_3CH_2C(O)H$. The other easily identifiable primary product was the corresponding fluorotelomer carboxylic acid (FTCA, $CF_3(CF_2)_xCH_2C(O)OH$). Another primary product was formed, for which standards were not available, but this was determined to be the corresponding fluorotelomer carboxylic peracid ($CF_3(CF_2)_xCH_2C(O)OOH$). There have not been any specific studies of reaction (60) in the presence of NO_x, but indirect evidence suggests that the mechanism and products are different under these conditions (Sulbaek Andersen et al. 2005a). The atmospheric fate of FTALs is expected to be dominated by reaction with hydroxyl radical (reaction (61)), but the products of this reaction have not been studied. Since the reaction is initiated by hydrogen abstraction, the products of chlorine atom-initiated oxidation are a reasonable proxy for those expected from reaction with hydroxyl radicals. Thus, under low-NO_x conditions, products of reaction (61) would consist of the corresponding PFAL,

FTCA, and fluorotelomer carboxylic peracid. The products in the presence of NO_x require further study.

The products of photolysis of FTALs (reaction (62)) have not been studied, but by analogy to PFALs and hydrocarbon aldehydes, it is likely that photolysis occurs via C–C bond scission:

$$CF_3(CF_2)_xCH_2C(O)H + h\nu \rightarrow CF_3(CF_2)_xC(^\bullet)H_2 + HCO \qquad (63)$$

The first stable product of the radical formed in reaction (63) would be the corresponding PFAL.

Through photolysis and hydroxyl radical-initiated atmospheric oxidation, FTALs form PFALs, which are known PFCA precursors. Consequently, FTALs can be considered sources of PFCAs.

3.4.4 Odd Fluorotelomer Alcohols (oFTOHs)

Atmospheric Lifetime

The atmospheric lifetime of odd carbon-numbered fluorotelomer alcohols (oFTOHs, $CF_3(CF_2)_xCH_2OH$) may be limited by reaction with chlorine atoms and hydroxyl radicals:

$$CF_3(CF_2)_xCH_2OH + Cl \rightarrow products \qquad (64)$$

$$CF_3(CF_2)_xCH_2OH + OH \rightarrow products \qquad (65)$$

The kinetics of reaction (64) have been studied indirectly using relative rate techniques (Hurley et al. 2004b) and directly using very low-pressure reactor mass spectrometry (Papadimitriou et al. 2007) (Table 7). Hurley et al. (2004b) did not observe any reactivity difference between chain lengths and cited a final rate constant of $k\,(Cl + CF_3(CF_2)_xCH_2OH) = (6.48 \pm 0.53) \times 10^{-13}\,cm^3/molecule/s$. The rate constant determined by Papadimitriou et al. (2007) is in good agreement with this value. Reaction (65) has been the subject of two studies (Chen et al. 2000; Hurley et al. 2004b) (Table 8). As with reaction (64), there does not appear to be an observable chain-length effect on the kinetics of reaction with hydroxyl radical. Given the low abundance of chlorine atoms in the atmosphere (Singh et al. 1996), the atmospheric lifetime of oFTOHs is dominated by reaction with hydroxyl radicals. Assuming an average atmospheric concentration of hydroxyl radicals of 1.1×10^6 molecules/cm^3, a lifetime of 124 d was established by Chen et al. (2000). Hurley et al. (2004b) determined a lifetime of 164 d for oFTOHs by comparing it to the reaction of CH_3CCl_3 with hydroxyl radicals.

Products of Atmospheric Degradation

The products of reaction (64) under low-NO_x conditions were studied using smog chamber techniques with Fourier-transform infrared spectroscopy (FTIR) detection.

The sole primary product from reaction of oFTOHs of various chain lengths was observed to be the corresponding PFAL (Hurley et al. 2004b). It is likely that hydrogen abstraction from oFTOHs yields PFALs as the primary product, when initiated by both chlorine atoms and hydroxyl radicals; the atmospheric fate of oFTOHs can be described as:

$$CF_3(CF_2)_xCH_2OH + Cl/OH \rightarrow CF_3(CF_2)_xC(O)H + HCl/H_2O \qquad (66)$$

The PFALs are known precursors of PFCAs, as described in Section 3.1.1. In addition, the formation of $CF_3C(O)OH$ was observed in situ by FTIR from the chlorine atom-initiated oxidation of CF_3CH_2OH (Hurley et al. 2004b), confirming that oFTOHs are PFCA precursors.

3.4.5 Even Fluorotelomer Alcohols (FTOHs)

Atmospheric Lifetime

The atmospheric fate of even carbon-numbered fluorotelomer alcohols ($CF_3(CF_2)_xCH_2CH_2OH$, FTOHs) has been the subject of numerous studies. FTOHs can react with atmospheric oxidants:

$$CF_3(CF_2)_xCH_2CH_2OH + Cl \rightarrow products \qquad (67)$$

$$CF_3(CF_2)_xCH_2CH_2OH + OH \rightarrow products \qquad (68)$$

Reaction (67) has been studied for $CF_3CH_2CH_2OH$ using relative rate techniques (Hurley et al. 2005; Kelly et al. 2005) and has been measured directly using very low-pressure reactor mass spectrometry (Papadimitriou et al. 2007). As shown in Table 7, the rates reported by Kelly et al. (2005) appear to be unusually high and the reasons for this discrepancy are unclear. The rates for longer-chained FTOHs ($CF_3(CF_2)_xCH_2CH_2OH$, $x = 3, 5, 7$) were studied using relative rate techniques (Ellis et al. 2003). Apart from the anomalous value, measurements of the reaction of FTOHs with chlorine atoms are in good agreement and appear to be independent of chain length. Thus, the rate of reaction of FTOHs with chlorine atoms is k (Cl + $CF_3(CF_2)_xCH_2CH_2OH$) = $(1.59 - 1.90) \times 10^{-11}$ cm^3/molecule/s. The kinetics of reaction (68) for $CF_3CH_2CH_2OH$ have been studied using relative rate techniques (Hurley et al. 2005; Kelly et al. 2005) and pulsed laser photolysis–laser-induced fluorescence (Kelly et al. 2005). As shown in Table 7, values are reasonably similar, but are not within experimental error of each other. The cause of the inconsistency is not clear. Kinetics for the longer-chained FTOHs ($CF_3(CF_2)_xCH_2CH_2OH$, $x = 3, 5, 7$) have been determined using relative rates (Ellis et al. 2003; Kelly et al. 2005), and the study results are in agreement. Chain length does not appear to affect the kinetics of reaction with hydroxyl radicals, although measurements are not all within experimental error. Taking into account the values in Table 8, the observed range for rate of reaction of FTOHs with hydroxyl radicals is $k\,(OH + CF_3(CF_2)_xCH_2OH) = (0.69 - 1.07) \times 10^{-12}\,cm^3/molecule/s$.

The low abundance of chlorine atoms in the atmosphere (Singh et al. 1996) suggests that reaction with chlorine will not be a dominant sink for FTOHs. Ellis et al. (2003) scaled the rate of the FTOH reaction with hydroxyl radicals to the lifetime of CH_3CCl_3 and determined an overall rate of approximately 20 d. Using a globally averaged concentration of hydroxyl radicals of 1×10^6 molecules/cm^3, a lifetime for FTOHs of about 12 d was estimated, which may be an underestimate due to the anomalously high rate constant (Kelly et al. 2005). Saturated compounds, such as FTOHs, do not react appreciably with ozone. Hurley et al. (2005) confirmed that $CF_3CH_2CH_2OH$ was unreactive with ozone and determined a minimum lifetime of 5,900 d, assuming a background ozone concentration of 40 ppb. Photolysis of FTOHs is also unlikely, because alcohols typically do not absorb within the actinic spectrum of the lower atmosphere. This was confirmed by determining the UV absorption of FTOHs using computational methods (Waterland et al. 2005), with which absorption was demonstrated to occur in the region of 140–175 nm, but with no absorption in the actinic region. In addition, the potential for wet and dry deposition was examined using simple expressions and lifetimes for $CF_3(CF_2)_7CH_2CH_2OH$, resulting in values of 2.5×10^6 and 8.4 years, respectively (Ellis et al. 2003). Consequently, the results of experiments suggest that the atmospheric fate of FTOHs is limited by reaction with hydroxyl radicals and have lifetimes on the order of 10–20 d. Atmospheric residence times have also been determined using both measured atmospheric concentrations of FTOHs and the Junge method (Dreyer et al. 2009b; Piekarz et al. 2007). The atmospheric residence times are in agreement with the experimental lifetime cited above, within the error of the methods used.

Products of Atmospheric Degradation

The observation that chlorine atom-initiated atmospheric oxidation of FTOHs results in PFCA production (Ellis et al. 2004) led to in-depth research into the underlying mechanism and into the overall atmospheric fate of FTOHs and intermediate species; results of much of this research have been addressed in previous sections. As a consequence, the products of atmospheric oxidation of FTOHs are well understood.

The corresponding FTAL has been observed in several studies to be the sole primary product resulting from the reaction of FTOHs (of various chain lengths) with chlorine atoms in the absence of NO$_x$ (Chiappero et al. 2008; Ellis et al. 2004; Hurley et al. 2004a, 2005; Kelly et al. 2005; Papadimitriou et al. 2007). The same trend has also been observed for the reaction of FTOHs with hydroxyl radicals in the absence of NO$_x$ (Kelly et al. 2005). The formation of the FTAL suggests that hydrogen abstraction occurs at the carbon adjacent to the alcohol group. This is corroborated by structure–activity relationship calculations (Ellis et al. 2003), which indicate that greater than 90% of reaction occurs at the alcohol-adjacent group, and by computational studies (Papadimitriou et al. 2007) that show the weakest C–H bonds are adjacent to the alcohol.

The formation of FTALs from FTOHs is thought to occur through the initial formation of an α-hydroxyl alkyl radical that reacts with molecular oxygen to form a chemically excited peroxy radical:

$$CF_3(CF_2)_xCH_2CH_2OH + Cl/OH \rightarrow CF_3(CF_2)_xCH_2C(^\bullet)HOH + HCl/H_2O \quad (69)$$

$$CF_3(CF_2)_xCH_2C(^\bullet)HOH + O_2 \rightarrow [CF_3(CF_2)CH_2C(OO^\bullet)HOH]^* \quad (70)$$

The radical product of reaction (70) can either collisionally deactivate or can rapidly decompose:

$$[CF_3(CF_2)_xCH_2C(OO^\bullet)HOH]^* + M \rightarrow CF_3(CF_2)_xCH_2C(OO^\bullet)HOH + M \quad (71a)$$

$$[CF_3(CF_2)_xCH_2C(OO^\bullet)HOH]^* \rightarrow CF_3(CF_2)_xCH_2C(O)H + HO_2 \quad (71b)$$

Decomposition typically occurs on a timescale faster than deactivation and yields the corresponding FTAL and HO_2. Collisionally deactivated peroxy radicals have two possible fates: decomposition to yield FTAL and HO_2 or reaction with NO_x:

$$CF_3(CF_2)_xCH_2C(OO^\bullet)HOH \rightarrow CF_3(CF_2)_xCH_2C(O)H + HO_2 \quad (72a)$$

$$CF_3(CF_2)_xCH_2C(OO^\bullet)HOH + NO \rightarrow products \quad (72b)$$

In the absence of NO_x, reaction (72b) does not occur and the sole product is the FTAL. In smog chamber experiments performed in the presence of abundant NO_x, FTAL was not the exclusive primary product (Hurley et al. 2005; Sulbaek Andersen et al. 2005a). However, when rates were scaled to more realistic NO_x concentrations of [NO] < 40 ppb, which encompass all but highly polluted urban conditions, it was determined that reaction (72b) would be of minimal importance. This was supported by further studies (Kelly et al. 2005), in which FTAL was observed to be the sole product of chlorine atom and hydroxyl radical-initiated atmospheric oxidation of FTOHs in the presence and absence of NO_x. Thus, the atmospheric fate of FTOHs can be understood as atmospheric oxidation through reactions (71b) and (71a) followed by reaction (72a), to form the corresponding FTAL:

$$CF_3(CF_2)_xCH_2CH_2OH + Cl/OH \rightarrow CF_3(CF_2)_xCH_2C(O)H \quad (73)$$

The FTALs are known precursors of PFCAs (Section *Perfluoroacyl Peroxy Radical Mechanism*), which suggests that PFCAs are formed from FTOHs. In addition, PFCAs have been directly observed in smog chamber experiments of the atmospheric oxidation of FTOHs, both by in situ FTIR (Hurley et al. 2004a) and offline sampling (Ellis et al. 2004); such experiments demonstrate their importance as precursors.

3.4.6 Fluorotelomer Olefins (FTOs)

Atmospheric Lifetime

The atmospheric reactivity of fluorotelomer olefins ($CF_3(CF_2)_xCH=CH_2$, FTOs) has been well documented. The lifetime of these compounds is limited by reaction with atmospheric oxidants:

$$CF_3(CF_2)_xCH=CH_2 + Cl \rightarrow products \qquad (74)$$

$$CF_3(CF_2)_xCH=CH_2 + OH \rightarrow products \qquad (75)$$

$$CF_3(CF_2)_xCH=CH_2 + O_3 \rightarrow products \qquad (76)$$

Rate constants for reaction (74) were determined for various FTO chain lengths ($CF_3(CF_2)_xCH=CH_2$, $x = 0, 1, 3, 5, 7$), using relative rate techniques at 296 K (Sulbaek Andersen et al. 2005b). There was no discernable effect of perfluorinated chain length on chlorine atom reactivity and a final rate constant for FTOs was cited as $k(Cl + CF_3(CF_2)_xCH=CH_2) = (9.07 \pm 1.08) \times 10^{-11}$ cm^3/molecule/s (Table 7). Studies of chlorine atom-initiated oxidation were also performed at low pressures and various temperatures using pulsed laser photolysis–resonance fluorescence (Vesine et al. 2000). Rate constants were shown to be independent of pressure and chain length for $CF_3(CF_2)_xCH=CH_2$ ($x = 3, 5$) and were in good agreement with those measured at atmospheric pressure.

Reaction (75) has been the subject of three separate studies (Table 8) (Orkin et al. 1997; Sulbaek Andersen et al. 2005b; Vesine et al. 2000). A study performed under smog chamber conditions at 296 K demonstrated that perfluorinated chain length did not affect the rate for the reaction of FTOs with hydroxyl radicals and a final rate of $k(OH + CF_3(CF_2)_xCH=CH_2) = (1.36 \pm 0.25) \times 10^{-12}$ cm^3/molecule/s was determined (Sulbaek Andersen et al. 2005b). This rate is in good agreement with a direct measurement for $CF_3(CF_2)_5CH=CH_2$ at 298 K (Vesine et al. 2000). A measurement for $CF_3CH=CH_2$ under flash photolysis–resonance fluorescence conditions (Orkin et al. 1997) is slightly higher than rate constants from the other two studies and may indicate a slight chain-length effect on moving from $CF_3CH=CH_2$ to FTOs with longer perfluorinated tails.

Studies of reaction (76) showed that the rate of reaction for $CF_3CH=CH_2$ with ozone was much faster than for $CF_3CF_2CH=CH_2$, but no further decrease in rate was observed with increasing chain length. Rate constants were, respectively, determined to be $k(O_3 + CF_3CH=CH_2) = (3.5 \pm 0.25) \times 10^{-19}$ cm^3/molecule/s and $k(O_3 + CF_3(CF_2)_xCH=CH_2, x \geq 1) = (2.0 \pm 0.4) \times 10^{-19}$ cm^3/molecule/s.

Atmospheric concentrations of chlorine atoms are too low (Singh et al. 1996) to impact the fate of FTOs. Assuming a global average hydroxyl radical concentration of 1×10^6 molecules/cm^3, the atmospheric lifetime of FTOs with respect to reaction with hydroxyl radicals is about 8.5 d. Using the global background concentration of ozone of 35 ppb, the lifetime of FTOs ($CF_3(CF_2)_xCH=CH_2$, $x \geq 1$) for reaction with ozone is about 70 d. Thus, the overall atmospheric lifetime of FTOs is 7.6 d, with hydroxyl radical reactions contributing 90% and ozone reactions contributing about 10% (Sulbaek Andersen et al. 2005b).

Products of Atmospheric Degradation

The products of FTO degradation via reactions (33) and (75) have been examined in two detailed studies (Nakayama et al. 2007; Vesine et al. 2000).

Initial studies of FTO reaction with chlorine atoms demonstrated the formation of a carbonyl-containing primary product that was tentatively identified as $CF_3(CF_2)_xC(O)CH_2Cl$ ($x = 3, 5$) (Vesine et al. 2000). The production of $CF_3C(O)CH_2Cl$ was confirmed from $CF_3CH=CH_2$, and the yield was quantified as $70 \pm 5\%$ in the absence of NO_x (Nakayama et al. 2007). A small amount of $CF_3C(O)H$ was also observed as a primary product with a yield of $6.2 \pm 0.5\%$. Using the yields of these two primary products, which occur from a reaction at the internal carbon atom, it was determined that chlorine atom addition occurs 74% to the terminal- and 26% to the internal-carbon atoms (Nakayama et al. 2007). Secondary reaction products, including COF_2 and CO, were observed and attributed to degradation of the primary carbonyl products (Vesine et al. 2000). In the presence of NO_x, chlorine atom-initiated reaction was observed to form additional nitrogen-containing species. In short experiments with $CF_3CH=CH_2$, the production of two primary species, identified as a nitrite ($CF_3C(ONO)HCH_2Cl$) and nitrate ($CF_3C(ONO_2)HCH_2Cl$), was observed (Nakayama et al. 2007). Experiments of greater duration with the longer-chain FTOs ($CF_3(CF_2)_xC(O)CH_2Cl$ ($x = 3, 5$)) also led to the production of peroxyacyl nitrate (PAN)-like species, $CF_3(CF_2)_xC(O)O_2NO_2$, which are likely formed as secondary products from degradation of the primary carbonyl species (Vesine et al. 2000). However, NO_x concentrations used in the studies were much higher than typical atmospheric concentrations and resulting yields of NO_x-containing compounds were likely inflated.

The reaction of FTO with hydroxyl radicals has been shown to yield the corresponding PFAL as the primary product for $CF_3(CF_2)_xC=CH_2$ ($x = 0, 3, 5$) (Nakayama et al. 2007; Vesine et al. 2000). A yield of $88 \pm 9\%$ was determined for PFALs from FTOs ($CF_3(CF_2)_xC=CH_2$ ($x = 0, 3$) (Nakayama et al. 2007). The only other carbon-containing species observed was a trace amount of COF_2, which was attributed to the degradation of PFALs. The production of PFALs from FTOs in a yield close to unity suggests that the fate of FTOs with hydroxyl radicals can be represented as:

$$CF_3(CF_2)_xCH=CH_2 + OH \rightarrow CF_3(CF_2)C(O)H \qquad (77)$$

There have not been any studies in which the fate of FTOs with ozone has been examined (reaction (76)). However, given that atmospheric oxidation of FTOs is dominated by reaction with hydroxyl radicals, the principal product of FTO degradation is the corresponding PFAL. As such, the atmospheric oxidation of FTOs could be a source of PFCAs.

3.4.7 Fluorotelomer Iodides (FTIs)

Atmospheric Lifetime

There is only one study that addresses the atmospheric fate of fluorotelomer iodides ($CF_3(CF_2)_xCH_2CH_2I$, FTIs), and it focuses on a representative compound, $C_4F_9CH_2CH_2I$ (Young et al. 2008). There are three potential pathways for the atmospheric fate of FTIs, including reaction with chlorine atoms or hydroxyl radicals:

$$C_4F_9CH_2CH_2I + Cl \rightarrow \text{products} \tag{78}$$

$$C_3F_9CH_2CH_2I + OH \rightarrow \text{products} \tag{79}$$

As with other iodinated species, FTIs are also probable subjects for photolysis:

$$C_4F_9CH_2CH_2I + h\nu \rightarrow \text{products} \tag{80}$$

Rate constants for reactions (78) and (79) were determined using competition kinetics at 295 K. The rate constant for reaction with chlorine is $k(Cl + C_4F_9CH_2CH_2I) = (1.25 \pm 0.15) \times 10^{-12}$ cm^3/molecule/s (Table 7), while that for reaction with hydroxyl radical is $k(OH + C_4F_9CH_2I) = (1.2 \pm 0.6) \times 10^{-12}$ cm^3/molecule/s (Table 8) (Young et al. 2008). Kinetics of reaction for fluorotelomer species have been shown to be independent of perfluorinated chain length (Hurley et al. 2004b, c), so rate constants for longer-chain-length FTIs are probably similar.

Average atmospheric concentrations of chlorine atoms are too low to impact the lifetime of FTIs (Singh et al. 1996). Using a globally averaged concentration of hydroxyl radicals of 1×10^6 molecules/cm^3 gives an approximate atmospheric lifetime with respect to reaction with OH of about 10 d. The susceptibility of FTIs to photolysis was determined using the TUV model (Young et al. 2008). The gas-phase UV–visible absorption cross-section of $C_4F_9CH_2CH_2I$ was measured and a quantum yield of unity was assumed, which is consistent with other alkyl iodides. The atmospheric lifetime with respect to photolysis was determined to be about 1 d at 44°N latitude. Overall, the photolysis lifetime is expected to range from about 1 d to 1 week, depending on time of year and latitude. This lifetime should be similar for FTIs of all chain lengths, because UV–visible cross-sections have minor dependence on perfluorinated chain length. Photolysis occurs on a shorter timescale than reaction with oxidants, indicating the lifetime of FTIs is limited by photolysis and will be approximately 1–7 d.

Products of Atmospheric Degradation

Products of the atmospheric oxidation of $C_4F_9CH_2CH_2I$ with chlorine radicals were studied using smog chamber techniques, with in situ detection by FTIR and offline detection by LC–MS–MS (Young et al. 2008). In the absence of NO_x, products detected by FTIR accounted for 59% of the loss of 4:2 FTI and were identified as $C_4F_9CH_2C(O)OOH$, $C_4F_9CH_2C(O)H$, COF_2, $C_4F_9C(O)H$, $C_4F_9C(O)OH$ (per-fluoropentanoic acid; PFPeA), and CO. Evidence of an unknown product was also observed, which could be the product $C_4F_9CH_2CH_2Cl$ or other products of reactive iodine species that are formed in the chamber. In offline product studies, PFCAs $(CF_3(CF_2)_xC(O)OH, \ x = 0 - 3)$ were identified as minor products of the oxidation of 4:2 FTI, which is consistent with the observed formation of PFPeA and known PFCA precursors in the smog chamber.

The products of photolysis of $C_4F_9CH_2CH_2I$ were also examined under smog chamber conditions (Young et al. 2008). The dominant product observed by FTIR, following photolysis of $C_4F_9CH_2CH_2I$ in the absence of NO_x, was the corresponding FTAL, suggesting it is produced as a primary product. This is expected to occur for all chain lengths of FTIs:

$$C_xF_{2x+1}CH_2CH_2I + h\nu \rightarrow C_xF_{2x+1}CH_2C(O)H \tag{81}$$

Reaction of FTIs with atmospheric oxidants has been shown to form PFCAs or known PFCA precursors. The dominant fate of FTIs is photolysis, which yields FTAL as a primary product and is a recognized PFCA precursor (Section 3.4.2). Thus, FTIs are potential sources of PFCAs to the environment.

3.4.8 Fluorotelomer Acrylate (FTAc)

Atmospheric Lifetime

The atmospheric fate of fluorotelomer acrylates (FTAcs, $CF_3(CF_2)_xCH_2CH_2OC(O)CH=CH_2$) has been examined in one study, focusing on a representative compound, $C_4F_9CH_2CH_2OC(O)CH=CH_2$ (Butt et al. 2009). The likely fate of this chemical is its reaction with chlorine atoms, hydroxyl radicals, or ozone:

$$C_4F_9CH_2CH_2OC(O)CH=CH_2 + Cl \rightarrow products \tag{82}$$

$$C_4F_9CH_2CH_2OC(O)CH=CH_2 + OH \rightarrow products \tag{83}$$

$$C_4F_9CH_2CH_2OC(O)CH=CH_2 + O_3 \rightarrow products \tag{84}$$

Rate constants for reactions (82) and (83) were determined using competition kinetics at 296 K (Butt et al. 2009). The rate for the reaction with chlorine is $k(Cl + C_4F_9CH_2CH_2OC(O)CH=CH_2) = (2.21 \pm 0.16) \times 10^{-10} \ cm^3/molecule/s$ (Table 7), while that for the reaction with hydroxyl radical is $k(OH + C_4F_9CH_2CH_2OC(O)CH=CH_2) = (1.13 \pm 0.12) \times 10^{-11} \ cm^3/molecule/s$ (Table 8)

(Butt et al. 2009). Chlorine atoms and hydroxyl radicals are presumed to react primarily at the double bond, and this is consistent with the observed rate constants. Rate constants for reaction (83) were indistinguishable within experimental error from non-fluorinated acrylates, such as methyl acrylate, suggesting the fluorinated tail does not impact the reactivity. No measurements to determine the rate of reaction with ozone were performed, but by comparison to methyl methacrylate, reaction (84) can be estimated to be $k(O_3 + C_4F_9CH_2CH_2OC(O)CH{=}CH_2) = 1 \times 10^{-18}$ cm^3/molecule/s. The kinetics of reaction for fluorinated species have been shown to be independent of perfluorinated chain length (Hurley et al. 2004b, c); therefore, the rate constants determined for $C_4F_9CH_2CH_2OC(O)CH{=}CH_2$ should apply to FTAcs of all chain lengths.

The concentrations of chlorine atoms in the atmosphere are not sufficient to impact the lifetime of FTAcs (Singh et al. 1996). Assuming a hydroxyl radical concentration of 1×10^6 molecules/cm^3, an atmospheric lifetime of FTAcs of approximately 1 d was determined. Assuming a concentration of 50 ppb for ozone yields a lifetime of approximately 9 d. Thus, the atmospheric lifetime of FTAcs are limited by reaction with hydroxyl radical and are on the order of 1 d.

Products of Atmospheric Degradation

Products of chlorine atom-initiated degradation of $C_4F_9CH_2CH_2OC(O)CH{=}CH_2$ were studied using FTIR under smog chamber conditions, in the presence and absence of NO$_x$ (Butt et al. 2009). Evidence of the production of $C_4F_9CH_2C(O)H$ as a primary product was observed, with yields of 10 and 18% in the presence and absence of NO$_x$, respectively. Although reaction of chlorine atoms occurs primarily by addition at the double bond, this suggests the occurrence of hydrogen abstraction as an additional loss mechanism:

$$C_4F_9CH_2CH_2OC(O)CH{=}CH_2 + Cl \rightarrow C_4F_9CH_2C(O)H + {}^{\bullet}OC(O)CH{=}CH_2 + HCl \tag{85}$$

Other FTIR features that could not be explicitly identified were attributed to a single product, $C_4F_9CH_2CH_2OC(O)C(O)HCH_2Cl$, that was formed from the addition of chlorine to the double bond:

$$C_4F_9CH_2CH_2OC(O)CH{=}CH_2 + Cl \rightarrow C_4F_9CH_2CH_2OC(O)C(O)HCH_2Cl \tag{86}$$

A single product resulted from the reaction of $C_4F_9CH_2CH_2OC(O)CH{=}CH_2$ with hydroxyl radicals, corresponding to the fluorotelomer glyoxalate (FTGly; $C_4F_9CH_2CH_2OC(O)CHO$). Perfluorinated chain length is not expected to affect the products of atmospheric oxidation, hence, the reaction of FTAcs with hydroxyl radicals can be described by the following reaction:

$$CF_3(CF_2)_xCH_2CH_2OC(O)CH{=}CH_2 + OH \rightarrow CF_3(CF_2)_xCH_2CH_2OC(O)CHO + HCHO \tag{87}$$

The atmospheric fate of FTGlys was not studied explicitly, but was suggested to be dominated by photolysis. The chromophores in FTGlys are similar to those in $CH_3C(O)C(O)H$, which has an atmospheric lifetime with respect to photolysis of about 3 h at a latitude of 45°. Thus, FTGlys probably degrade via photolysis on a similar timescale. By analogy to $CH_3C(O)C(O)H$, photolysis is expected to break the carbon–carbon bond between the dicarbonyl $(CF_3(CF_2)_xCH_2CH_2OC(O)-C(O)H)$, which would ultimately lead to the formation of FTAL:

$$CF_3(CF_2)_xCH_2CH_2OC(O)C(O)H + h\nu \rightarrow CF_3(CF_2)_xCH_2C(O)H \qquad (88)$$

These data suggest that the fluorotelomer aldehyde, $CF_3(CF_2)_xCH_2C(O)H$, is formed as the secondary product of atmospheric oxidation of FTAcs (Butt et al. 2009). The FTAL then reacts as described in Section 3.4.2 to form PFCAs in small yields. Consequently, FTAcs are potential sources of PFCAs to the environment.

3.5 Perfluoroalkanesulfonamides

Many perfluoroalkanesulfonamides of varied structures were used commercially. Prior to 2001, eight-carbon congeners were the dominant products, while more recently four-carbon congeners have been primarily produced.

3.5.1 N-Alkyl-perfluoroalkanesulfonamides (NAFSA)

Atmospheric Lifetime

The atmospheric lifetime of a representative compound for this class, N-ethyl-perfluorobutanesulfonamide (NEtFBSA, $C_4F_9SO_2N(H)CH_2CH_3$), has been studied by Martin et al. (2006). The fate of this compound is governed by reaction with chlorine atoms and hydroxyl radicals:

$$C_4F_9SO_2N(H)CH_2CH_3 + Cl \rightarrow products \qquad (89)$$

$$C_5F_9SO_2N(H)CH_2CH_3 + OH \rightarrow products \qquad (90)$$

Kinetics of reactions (89) and (90) were examined using competition kinetics, with FTIR and offline LC–MS–MS detection. The rate constant for reaction with chlorine atoms was determined as $k(Cl + C_4F_9SO_2N(H)CH_2CH_3) = (8.37 \pm 1.44) \times 10^{-12}$ cm^3/molecule/s and the rate constant for reaction with hydroxyl radicals was measured to be $k(OH + C_4F_9SO_2N(H)CH_2CH_3) = (3.74 \pm 0.77) \times 10^{-13}$ cm^3/molecule/s (Table 11) (Martin et al. 2006).

Since atmospheric concentrations of chlorine atoms are very low (Singh et al. 1996), the gas-phase atmospheric lifetime is expected to be dominated by reaction with hydroxyl radicals. Reactions are also likely independent of perfluorinated chain length (Hurley et al. 2004b, c). Assuming an average OH

Table 11 Summary of chlorine atom- and hydroxyl radical-initiated kinetics for perfluorosulfonamides

	Structure	x	Rate constant (cm^3/molecule/s)	T (K)	Method	References
Cl	NEtFSA $CF_3(CF_2)_xSO_2N(H)CH_2CH_3$	3	$(8.37 \pm 1.44) \times 10^{-12}$	296	Relative rate	Martin et al. (2006)
OH	NEtFSA $CF_3(CF_2)_xSO_2N(H)CH_2CH_3$	3	$(3.74 \pm 0.77) \times 10^{-13}$	301	Relative rate	Martin et al. (2006)
	NAFSE $CF_3(CF_2)_xSO_2N(CH_3)CH_2CH_3$	3	$(5.8 \pm 0.8) \times 10^{-12}$	296	Relative rate	D'eon et al. (2006)

concentration of 1×10^6 molecules/cm^3, the lifetime for NEtFSAs are 20–50 d, depending on location, time of year, and temperature. It is important to note that the rate constants and lifetimes determined in this study are for gas-phase reactions only. The low volatility of these compounds suggests that gas–particle partitioning could play a role in the overall atmospheric fate of NAFSAs; however, the degree of partitioning remains unclear (Shoeib et al. 2004; Stock 2007). The effect of altering substitution at the nitrogen on atmospheric kinetics is unclear. Atmospheric residence times were determined for N-methyl-perfluorobutanesulfonamide (NMeFBSA), N-methyl-perfluorooctanesulfonamide (NMeFOSA), N-ethyl-perfluorooctanesulfonamide (NEtFOSA), and perfluorooctanesulfonamide (PFOSA) and were based on measured atmospheric concentrations using the Junge method as well as assumptions regarding partitioning (Dreyer et al. 2009b). Residence times were in reasonable agreement with the experimentally determined atmospheric lifetime for NEtFBSA of Martin et al. (2006).

Products of Atmospheric Degradation

The products of the reaction of NEtFBSA with chlorine atoms have been studied experimentally (Martin et al. 2006) and theoretically (Antoniotti et al. 2008). Experimental results demonstrated the formation of two identifiable primary stable products, formed from reaction at the secondary (reaction (91)) and primary (reaction (92)) carbons on the ethane moiety (Martin et al. 2006):

$$C_4F_9SO_2N(H)CH_2CH_3 + Cl \rightarrow C_4F_9SO_2N(H)C(O)CH_3 \tag{91}$$

$$C_4F_9SO_2N(H)CH_2CH_3 + Cl \rightarrow C_4F_9SO_2N(H)CH_2C(O)H \tag{92}$$

Two additional partially resolved peaks were observed in LC–MS–MS chromatograms of reacted NEtFBSA. These were assumed to be additional primary products, but were not unambiguously identified. Theoretical study results indicate that these products are a consequence of the hydrogen abstraction from the nitrogen atom and go on to form one of three isomeric stable products (Antoniotti et al. 2008):

$$C_4F_9SO_2N(H)CH_2CH_3 + Cl \rightarrow C_4F_9SO_2N(OH)CH_2CH_2 \tag{93a}$$

$$C_4F_9SO_2N(H)CH_2CH_3 + Cl \rightarrow C_4F_9SO_2ON(H)CH_2CH_2 \tag{93b}$$

$$C_4F_9SO_2N(H)CH_2CH_3 + Cl \rightarrow C_4F_9SO_2N(H)OCH_2CH_2 \tag{93c}$$

It is suggested that the product of reaction (93a) is dominant, with products of reactions (93b) and (93c) playing a minor role, if formed at all. Following reactions in the smog chamber, COF$_2$ and SO$_2$ were observed by FTIR, and the homologous series of PFCAs (CF$_3$(CF$_2$)$_x$C(O)OH, $x = 0 - 2$) were observed by LC–MS–MS (Martin et al. 2006). The PFCAs are likely formed from the perfluorobutyl radical, $CF_3CF_2CF_2CF_2^\bullet$, by the mechanism described in *Perfluorinated Radical Mechanism* Section. This radical could be created through degradation of the products of reactions (91) and (92) to form $C_4F_9SO_2^\bullet$, which loses SO_2 to yield the perfluorobutyl radical. Results of computational studies also suggest that products of reactions (93b) and (93c) can decompose into $C_4F_9SO_2^\bullet$ or directly into $C_4F_9^\bullet$ and SO_2 (Antoniotti et al. 2008). Although these studies suggest that NEtFSAs can act as PFCA precursors, it is unclear how the differing alkyl substitution in NAFSAs would affect the observed products.

3.5.2 *N*-Alkyl-perfluoroalkanesulfamidoethanols (NAFSE)

Atmospheric Lifetime

The atmospheric lifetime for a representative compound from this class, *N*-methyl perfluorobutane sulfamidoethanol (NMeFBSE, $C_4F_9SO_2N(CH_3)CH_2CH_2OH$), was determined using competition kinetics (D'eon et al. 2006). The lifetime of this chemical is limited by atmospheric oxidation initiated by hydroxyl radicals:

$$C_4F_9SO_2N(CH_3)CH_2CH_2OH + OH \rightarrow \text{products} \tag{94}$$

The rate of this reaction was determined as $k(OH + C_4F_9SO_2N(CH_3)CH_2CH_2OH) = (5.8 \pm 0.8) \times 10^{-12} \, cm^3/molecule/s$ (Table 11) and is probably applicable to NMeFSEs with longer perfluorinated chains. The similarity in reactivity of NMeFBSE and n-C_3H_7OH with hydroxyl radicals suggests that reaction takes place primarily on the $-CH_2CH_2OH$ portion of NMeFBSE. Because that reactivity occurs primarily on the ethanol moiety, altering the alkyl substitution is unlikely to affect kinetics, so this rate probably applies to all NAFSEs. Using a globally averaged concentration of hydroxyl radicals of 1×10^6 molecules/cm^3, the gas-phase atmospheric lifetime of NAFSEs is approximately 2 d. The relatively low volatility of NAFSEs suggests that gas–particle partitioning may play a role in the overall atmospheric fate of this class of compounds, but the degree to which this occurs is not well defined (Shoeib et al. 2004; Stock 2007). Dreyer et al. (2009b) determined atmospheric residence times for NMeFBSE and *N*-methylperfluorooctanesulfamido ethanol (NMeFOSE), using measured atmospheric concentrations along with the Junge method and assumptions regarding partitioning. The atmospheric lifetime for NMeFOSE is assumed to be the same as the experimentally determined atmospheric lifetime of NMeFBSE as reported by D'eon et al. (2006). It is interesting to note that the atmospheric residence time determined by Dreyer et al. (2009b) for NMeFOSE agrees with this atmospheric lifetime, while the residence time for NMeFBSE is approximately one order of magnitude higher than the experimentally determined value. The reasons for this discrepancy are unclear.

Products of Atmospheric Degradation

D'eon et al. (2006) studied the products of reaction (94) under smog chamber conditions, with in situ FTIR detection and offline LC–MS–MS and GC–MS analyses. An aldehyde was observed as a product formed early in the reaction:

$$C_4F_9SO_2N(CH_3)CH_2CH_3OH + OH \rightarrow C_4F_9SO_2N(CH_3)CH_2C(O)H \qquad (95a)$$

An N-dealkylation product was also observed:

$$C_4F_9SO_2N(CH_3)CH_2CH_3OH + OH \rightarrow C_4F_9SO_2N(H)CH_3$$

These types of dealkylation reactions have been observed in the gas phase (Woodrow et al. 1978), but there have not been any mechanistic explanations made to date. Offline samples taken after reactions with hydroxyl radicals showed the homologous series of PFCAs ($CF_3(CF_2)_xC(O)OH$, $x = 0 - 2$) and $C_4F_9SO_3H$ (PFBS). The mechanism proposed to explain these observations involves addition of the hydroxyl radical to the sulfone double bond and subsequent breakage of the C–S or C–N bond to form PFCAs or PFBS, respectively. These reactions are described in detail in Section 2.2.

Experiments to determine the products of NMeFBSE reaction with chlorine atoms were also undertaken (D'eon et al. 2006):

$$C_4F_9SO_2N(CH_3)CH_2CH_2OH + Cl \rightarrow products \qquad (96)$$

The products observed were similar to those observed for reaction with hydroxyl radical, including the N-dealkylation products ($C_4F_9SO_2N(H)CH_3$), PFCAs, and PFBS. Chlorine atoms are often used as surrogates for hydroxyl radicals when oxidation is initiated by hydrogen abstraction, as in reactions (95a) and (95b). It is surprising that PFBS was observed from the chlorine atom-initiated oxidation of NMeFBSE, as this cannot be explained by the mechanism described above and in Section 2.2. An alternative mechanism can be found in computational work on the reaction of chlorine atoms with NEtFBSA (reaction (89)), which is similar in structure to the N-dealkylation product formed from NMeFBSE in reaction (95b) (Antoniotti et al. 2008). Abstraction of a hydrogen atom from the nitrogen, followed by reaction with molecular oxygen can lead to insertion of oxygen into the S–O bond (see also reaction (93b)):

$$C_xF_{2x-1}SO_2N(H)R + Cl \rightarrow C_xF_{2x-1}SO_2N(^\bullet)R \qquad (97)$$

$$C_xF_{2x-1}SO_2N(^\bullet)R + O_2 \rightarrow C_xF_{2x-1}SO_2OONR \qquad (98)$$

This product may decompose to form the corresponding sulfonic radical:

$$C_xF_{2x-1}SO_2OONR + Cl \rightarrow C_xF_{2x-1}SO_2O^\bullet + {}^\bullet ONR \qquad (99)$$

The fate of this sulfonic radical would presumably be reaction with HO_2 to yield the sulfonic acid:

$$C_xF_{2x-1}SO_2O^{\bullet} + HO_2 \rightarrow C_xF_{2x-1}SO_2OH + O_2 \tag{100}$$

Although the activation energy of reaction (98) is high, the reaction overall is exoergetic in nature, suggesting that this may be a feasible mechanism, especially given the highly exoergetic nature of reaction (99) (Antoniotti et al. 2008). While NMeFSEs can be considered precursors to PFSAs and PFCAs, the impact of altering alkyl substitutions in NAFSEs on product distribution is unknown.

4 Atmospheric Sources and Levels

4.1 Volatile Fluorinated Anesthetics

Both halothane and isoflurane are gases and are commonly used as anesthetics. Exposure to these compounds by medical personnel is of concern (OSHA Directorate for Technical Support 2000), where operating room levels can be up to hundreds of parts-per-billion (Al-Ghanem et al. 2008). Such levels of release suggest that the compounds will be ventilated into the atmosphere unless specific precautions are taken to prevent their release. Despite the likelihood of their presence, no measurements have been made of halothane and isoflurane in the atmosphere.

4.2 Hydrochlorofluorocarbons (HCFCs)

4.2.1 Potential Sources to the Atmosphere

HCFCs have been used as replacements for chlorofluorocarbons (CFCs) because of the lower ozone-depletion potential of the former. Uses of these chemicals are predictably increasing as the Montreal Protocol and related documents come into effect (UNEP 2000). These compounds are gases at environmentally relevant temperatures and may enter the environment through intentional and fugitive emissions. HCFCs are also slated for phase out through the Montreal Protocol, with production directed to be frozen in the next few years (UNEP 2000).

4.2.2 Atmospheric Concentrations

Concentrations of HCFC-124 were below detection limits using the available contemporary methods in the late 1990s. However, more recently, HCFC-124 has been measured in the atmosphere at concentrations between 1.34 and 1.67 pptv (parts-per-trillion by volume) in the troposphere, with a growth rate of 0.06–0.35 pptv/year (Velders et al. 2005). The most recent measurements available for HCFC-123 suggest that it remains at the sub-pptv level (Velders et al. 2005). Measurements of HCFC-225ca have not been made.

4.3 Hydrofluorocarbons (HFCs, Non-telomer Based)

4.3.1 Saturated Hydrofluorocarbons (HFCs)

Potential Sources to the Atmosphere

Hydrofluorocarbons (HFCs) are primarily used as replacements for ozone-depleting CFCs, because they contain no chlorine and have no impact on stratospheric ozone. These compounds are gases employed in the coolant industry that can be released by fugitive emissions. The emission of HFC-134a has been observed at elevated concentrations in a road traffic tunnel, presumably as a result of release from car air conditioners (Stemmler et al. 2004). As CFCs are phased out by the Montreal Protocol (UNEP 2000), production of HFCs has been increasing to compensate. However, there is concern about the ability of HFCs to act as long-lived greenhouse gases (Forster et al. 2007), which in some cases has prompted directives to eliminate their use. Legislation to replace HFC-134a as the major coolant in mobile air conditioners has been introduced by the European Union (The European Parliament and the Council of the European Union 2006), suggesting that production of this compound is likely to decrease in the coming years.

Atmospheric Concentrations

Concentrations of saturated HFCs are expected to be well mixed in the atmosphere due to their long lifetimes. Few measurements of HFC-125 have been made, but levels of 1.4–5.1 pptv have been detected at diverse locations around the world (Miller et al. 2008; Reimann et al. 2004; Velders et al. 2005). Consistent with expectations regarding usage, atmospheric concentrations of HFC-125 were observed to be increasing at a rate of 0.43–0.56 pptv/year (Reimann et al. 2004; Velders et al. 2005).

Atmospheric levels of HFC-134a have received a great deal of attention as a result of its high usage (Table 12). Levels have been steadily increasing from sub-pptv levels, in the late 1980s, to tens of pptv in the late 1990s to early 2000s. Northern hemisphere concentrations appear to be higher than southern hemisphere concentrations, presumably because of the larger source region present in the north. It also appears that concentrations may be higher in urban areas, which is consistent with population-driven usage and emissions.

Concentrations of other saturated HFCs have not yet been detected in the atmosphere.

Table 12 Gas-phase atmospheric concentrations of hydrofluorocarbon-134a

Concentration (pptv)	Trend (pptv/year)	Location	Date collected	Method	References
<0.05		Pacific Cruise, NH average	May, Jun 1987	Flasks	Montzka et al. (1996)
<0.05		Pacific Cruise, SH average	May, Jun 1987	Flasks	
<0.05		Pacific Cruise, global average	May, Jun 1987	Flasks	
0.1		Cape Meares, Oregon	1988	Flasks	Culbertson et al. (2004)
0.2		Cape Meares, Oregon	1989	Flasks	
0.2		Cape Meares, Oregon	1990	Flasks	
0.01		Cape Grim, Tasmania	Jun 1990	Flasks	Oram et al. (1996)
0.3		Cape Meares, Oregon	1991	Flasks	Culbertson et al. (2004)
0.4		Cape Meares, Oregon	1992	Flasks	
0.4		Palmer Station, Alaska	1992	Flasks	
0.7		Cape Meares, Oregon	1993	Flasks	
0.12		Atlantic Ocean	Nov 1993	Flasks	Oram et al. (1996)
1.6		Cape Meares, Oregon	1994	Flasks	Culbertson et al. (2004)
0.6		Palmer Station, Antarctica	1994	Flasks	
0.62		Pacific Cruise, NH average	Jan, Feb 1994	Flasks	Montzka et al. (1996)
0.26		Pacific Cruise, SH average	Jan, Feb 1994	Flasks	
0.44		Pacific Cruise, global average	Jan, Feb 1994	Flasks	
1.22		Atlantic Cruise, NH average	Oct, Nov 1994	Flasks	
0.57		Atlantic Cruise, SH average	Oct, Nov 1994	Flasks	
0.90		Atlantic Cruise, global average	Oct, Nov 1994	Flasks	
0.37		Atlantic Ocean	Nov 1994	Flasks	Oram et al. (1996)
0.43		Cape Grim, Tasmania	Dec 1994	Flasks	
1.48		Mace Head, Ireland	Dec 1994	Flasks	
2.6		Cape Meares, Oregon	1995	Flasks	Culbertson et al. (2004)
2.4		Point Barrow, Alaska	1995	Flasks	
1.0		Palmer Station, Antarctica	1995	Flasks	

Table 12 (continued)

Concentration (pptv)	Trend (pptv/year)	Location	Date collected	Method	References
1.38		NH average[a]	Jan 1995	Flasks	Montzka et al. (1996)
0.79		SH average[a]	Jan 1995	Flasks	
1.08		Global average[a]	Jan 1995	Flasks	
1.66		NH average[a]	Apr 1995	Flasks	
0.84		SH average[a]	Apr 1995	Flasks	
1.25		Global average[a]	Apr 1995	Flasks	
2.03		NH average[a]	May 1995	Flasks	
1.00		SH average[a]	May 1995	Flasks	
1.52		Global average[a]	May 1995	Flasks	
2.33		NH average[a]	Jul 1995	Flasks	
1.13		SH average[a]	Jul 1995	Flasks	
1.73		Global average[a]	Jul 1995	Flasks	
1	0.64 ± 0.02	Cape Grim, Tasmania	Nov 1995	Flasks	Oram et al. (1996)
3.3		Cape Meares, Oregon	1996	Flasks	Culbertson et al. (2004)
3.7		Point Barrow, Alaska	1996	Flasks	
2.0		Palmer Station, Antarctica	1996	Flasks	
3.67	2.05 ± 0.01	Mace Head, Ireland	Jan 1996	In situ	Simmonds et al. (1998)
4.67 ± 0.10	2.07 ± 0.09	Mace Head, Ireland	Aug 1996	In situ	Prinn et al. (2000)
4.8		Cape Meares, Oregon	1997	Flasks	Culbertson et al. (2004)
5.3		Point Barrow, Alaska	1997	Flasks	
2.9		Palmer Station, Alaska	1997	Flasks	
5.6 ± 0.05	2.4 ± 0.2	Cape Grim, Tasmania	Jul 1998	In situ	Prinn et al. (2000)
16.41 ± 0.04	3.78 ± 0.03	Mace Head, Ireland	1998–2002 fitted	In situ	O'Doherty et al. (2004)
11.36 ± 0.01	3.12 ± 0.01	Cape Grim, Tasmania	1998–2002 fitted	In situ	
11.02 ± 0.03	2.99 ± 0.03	Cape Grim, Tasmania	1998–2002 fitted	Flasks	
15		Jungfraujoch, Switzerland	Jan 2000	In situ	Reimann et al. (2004)

Table 12 (continued)

Concentration (pptv)	Trend (pptv/year)	Location	Date collected	Method	References
36		Guangzhou City, China	2001	Flasks	Chan et al. (2006)
49		Panyu, China	2001	Flasks	
19		Dinghu Mountain, China	2001	Flasks	
16		Western Pacific	2001	Flasks	
27		Jungfraujoch, Switzerland	Dec 2002	In situ	Reimann et al. (2004)
45	4.0	Trinidad Head, California	Mid 2006	In situ	Miller et al. (2008)

[a]NH sites: Alert, Canada; Point Barrow, Alaska; Niwot Ridge, Colorado; Mauna Loa, Hawaii. SH sites: Cape Matatula, American Samoa and Cape Grim, Tasmania

4.3.2 Hydrofluoroolefins (HFOs)

HFOs are used in the synthesis of fluorinated polymers and are proposed as replacements for CFC replacements that have a high climate impact, such as saturated HFCs. These compounds are also known pyrolysis products of perfluoroalkyl ethers (Hoshino et al. 1996) and fluoropolymers (Ellis et al. 2001). The incineration of waste that contains these polymers could act as an important environmental source. Jordan and Frank (1999) determined an approximate yield of TFA of 200 t/year from European incineration of fluoropolymers and subsequent release of hexafluoropropene. Currently, HFOs for use as coolants are in the development stage and are not produced in large quantities for this purpose. There have been no measurements made of these compounds in the atmosphere to date.

4.4 Fluorotelomer Compounds

4.4.1 Potential Sources to the Atmosphere

Surfactant-based polyfluorinated chemicals, used for their stain-repellent properties, are typically synthesized by either the telomerization or electrochemical process. Fluorotelomer compounds are synthesized using telomerization, in which products are normally characterized by linear molecules with even numbers of carbon atoms, though odd carbon chain lengths are also possible. The initial product of telomerization is the perfluorinated iodide $(CF_3(CF_2)_xI)$, from which a number of volatile fluorotelomer chemicals, such as fluorotelomer iodides (FTIs), fluorotelomer alcohols (FTOHs), fluorotelomer olefins (FTOs), and fluorotelomer acrylates (FTAcs), are synthesized. The initial product of the perfluorinated iodide is the FTI, which is then used to synthesize FTOHs, leaving $\leq$2% FTI as residual and forming 2–5% FTO as a by-product (Prevedouros et al. 2006). These compounds can be incorporated directly into consumer products. Alternatively, FTOHs or FTIs can be used to synthesize FTAcs, which are used to create a type of polymer that constitutes >80% of the fluoropolymer market (Prevedouros et al. 2006). The use of FTOHs to form FTAcs results in 0.1–0.5% by weight of FTOH residuals, while use of FTIs as a reaction precursor to FTAcs yields 3–8% by weight of FTOs as by-products (Prevedouros et al. 2006). Approximately 0.4% by mass of residual FTI, FTO, FTOH, and FTAc was detected in a commercial fluoropolymer (Russell et al. 2008). The presence of unreacted FTOHs of multiple chain lengths was also observed at levels of a few percent by mass in several commercial fluorotelomer products (Dinglasan-Panlilio and Mabury 2006).

The presence of unreacted materials, or residuals, in commercial products could be a major source of PFA precursors to the atmosphere. This is corroborated by high levels of fluorotelomer chemicals in indoor air (Barber et al. 2007; Jahnke et al. 2007d; Shoeib et al. 2008; Strynar and Lindstrom 2008). Indoor concentrations were consistently at least 10 times greater than outdoor air, and in some environments indoor:outdoor air-level ratios exceeded one hundred (Barber et al. 2007; Jahnke et al. 2007d). In response to this potential source to the atmosphere, some producers of fluorotelomer products have committed to reducing the residuals present in their products (United States Environmental Protection Agency 2006).

4.4.2 Atmospheric Concentrations

Atmospheric concentrations of semi-volatile PFA precursors, such as FTOHs, FTOs, and FTAcs, have been widely measured. Because of their relatively short life-times (<1 month), they are not uniformly distributed in the atmosphere and spatial measurements gain importance. These measurements have all been made by collection of the chemicals on a sorbent, followed by solvent extraction and analysis by gas chromatography coupled to mass spectrometry (GC–MS). Methods by which these compounds have been measured were comprehensively reviewed elsewhere (Jahnke and Berger 2009).

FTOHs

Published measurements of atmospheric FTOHs are shown in Table 13. These compounds were first measured in 2001 by Martin et al. (2002). Measurements have since been made in a number of locations, but have not been of global scope. Most measurements were made in the Northern Hemisphere and have been concentrated in North America, Europe, and Japan. The primary focus of most analyses has been on the 6:2, 8:2, and 10:2 FTOHs ($CF_3(CF_2)_xCH_2CH_2OH$, $x = 5,7,9$, respectively), though some measurements of 4:2 and 12:2 FTOHs ($x = 3,11$) have also been made. The levels detected show large variability, but some clear trends emerge. Levels of 8:2 FTOH are almost always highest, followed by 6:2 FTOH in North America and Europe, and 10:2 FTOH in Japan. Levels of 4:2 and 12:2 FTOHs tend to be lowest. In general, urban > semi-urban > rural > remote concentrations, with some exceptions that are probably due to air mass source. This trend is not surprising, given that the source of these compounds appears to be directly related to human population density. The high range of concentrations observed, from sub-pg/m^3 to >1,000 pg/m^3, illustrates the importance of source regions and atmospheric transport in determining the environmental concentrations of these compounds.

FTOs

The FTOs have been measured in only a few campaigns. The higher volatility of these compounds with respect to FTOHs precludes them from being analyzed by the methods commonly used for semi-volatile fluorinated compounds. Sorbent-based collection with solvent extraction and subsequent evaporation has typically led to poor recoveries for FTOs, particularly the shorter-chain congeners (Piekarz et al. 2007). Regardless, the few measurements that are available are presented in Table 14 and show similar trends as those seen for FTOHs. Urban concentrations are the highest, while rural and remote measurements show low levels of FTOs. In

Table 13 Summary of published atmospheric concentrations of FTOHs (CF$_3$(CF$_2$)$_x$CH$_2$CH$_2$OH)

Concentration (pg/m³)										
$x = 3$ (4:2)	$x = 5$ (6:2)	$x = 7$ (8:2)	$x = 9$ (10:2)	$x = 11$ (12:2)	Phase	Location	n	Date collected	Method	References
Urban										
nd	87 (30–196)	55 (9–123)	32 (7–46)		g+p	Toronto, ON	4	Feb–Apr 2001	1	Martin et al. (2002)
	(61–100)[a]	(51–118)[a]	nd		g+p	Toronto, ON	3	Nov 2001	1	Stock et al. (2004)
	(30–55)[a]	(30–60)[a]	nd		g+p	Reno, NV	3	Nov 2001	1	Stock et al. (2004)
	nd	11 (nd–18)	nd		g+p	Winnipeg, MB	3	Nov 2001	1	Stock et al. (2004)
54 (22–117)	66 (33–149)	119 (62–275)	35 (16–93)		g	Hamburg	7	Apr, May 2005	1	Jahnke et al. (2007a)
nd	nd	<1.0	<0.7		p	Hamburg	7	Apr, May 2005	1	Jahnke et al. (2007a)
31 (32,29)	56 (55,56)	108 (106,110)	29 (28,29)		g+p	Hamburg	2	Apr 2005	1	Jahnke et al. (2007b)
3	187	237	65		g	Manchester, UK	1	Feb, Mar 2005	1	Barber et al. (2007)
2.1	1.8	5.8	2.7		p					
<16.0	61.2	237	61.8	20.9	g	Manchester, UK	1	Nov, Dec 2005	1	Barber et al. (2007)
<9.7	<0.1	<1.8	<0.5	<6.1	p					
	17.7	40.2	21.1		g	Toronto, ON	3	Mar 2006	1	Shoeib et al. (2006)
	(12.4–27.2)	(25.1–59.6)	(12.0–36.1)							
	0.31	0.71	1.09		p					
	(0.20–0.42)	(0.30–1.31)	(0.42–1.82)							

Table 13 (continued)

Concentration (pg/m³)

$x=3$ (4:2)	$x=5$ (6:2)	$x=7$ (8:2)	$x=9$ (10:2)	$x=11$ (12:2)	Phase	Location	n	Date collected	Method	References
	23 (16–36)	24 (17–39)	9.3 (8.5–9.8)		g+p	Hamburg	5	Apr 2006	2	Jahnke et al. (2007d)
	21 (15–26)	52 (37–60)	17 (13–23)		g+p	Hamburg	4	Apr 2006	1	Jahnke et al. (2007d)
	10[b] (nd–28)	352[b] (48–1743)	39[b] (nd–197)		g+p	Kyoto City, Japan	10	Oct–Dec 2006	3	Oono et al. (2008b)
	35[b] (27–44)	595[b] (199–999)	81[b] (45–143)		g+p	Osaka City, Japan	4	Sept–Dec 2006	3	Oono et al. (2008b)
	29[b] (nd–170)	1465[b] (310–4585)	173[b] (35–518)		g+p	Osaka City, Japan	10	Oct, Nov 2006	3	Oono et al. (2008b)
	nd	99.2[b] (<32–421)	14.5[b] (<17–67.0)		g	Keihan Area, Japan	15	Apr–Jun 2007	4	Oono et al. (2008a)
	18.3[b] (<15–768)	90.2[b] (<32–2466)	18.5[b] (<17–113)		g	Across Japan	18	Apr–Jun 2007	4	Oono et al. (2008a)
nd	13	130	29		g	Hamburg	2	Oct, Nov 2007	1	Dreyer and Ebinghaus (2009)
nd	nq	nd	nd		p					
Semi-urban										
nq	(nd–40)[a]	(50–205)[a]	nq		g+p	Griffin, GA	5	Nov 2001	1	Stock et al. (2004)
nd	(52–93)[a]	(48–87)[a]	nd		g+p	Cleves, OH	3	Nov 2001	1	Stock et al. (2004)

Table 13 (continued)

Concentration (pg/m³)										
$x = 3$ (4:2)	$x = 5$ (6:2)	$x = 7$ (8:2)	$x = 9$ (10:2)	$x = 11$ (12:2)	Phase	Location	n	Date collected	Method	References
56.5	81	102	75		g	Hazelrigg, UK	1	Feb, Mar 2005	1	Barber et al. (2007)
0.7	<1.1	<1.1	<1.1		p	Hazelrigg, UK	1	Feb, Mar 2005	1	Barber et al. (2007)
26.3	31.5	66.5	21.2	7.6	g	Hazelrigg, UK	1	Nov 2005–Feb 2006	1	Barber et al. (2007)
<9.7	<0.1	8.5	<0.5	<6.1	p					
nd	(16.1,26.3)	(96.7,104.3)	(27.7,36.8)	(7.0,12.4)	g+p	Hamburg	2	Jun 2006	1	Dreyer et al. (2008)
nd	(13–38)	(30–127)	(13–34)	(2.5–5.5)	g	Barsbuettel	3	Oct, Nov 2007	1	Dreyer and Ebinghaus (2009)
(nd–5.8)	(7–54)	(17–96)	(6–32)	(1.8–10)	g	Hamburg	5	Oct, Nov 2007	1	Dreyer and Ebinghaus (2009)
Rural										
nd	29 (41,16)	55 (40,25)	17 (20,15)		g+p	Long Point, ON	2	Feb–Apr 2001	1	Martin et al. (2002)
26 (nd–52)	nd	nd	nd		g+p	Long Point, ON	3	Nov 2001	1	Stock et al. (2004)

Table 13 (continued)

Concentration (pg/m³)					Phase	Location	n	Date collected	Method	References
$x = 3$ (4:2)	$x = 5$ (6:2)	$x = 7$ (8:2)	$x = 9$ (10:2)	$x = 11$ (12:2)						
19 (3.3–45) nd	64 (17–125) nd	75 (33–112) <1.0	23 (10–32) <0.7		g p	Waldhof	4	May, Jun 2005	1	Jahnke et al. (2007a)
9 (7,11)	29 (29,29)	85 (81,88)	28 (27,29)		g+p	Waldhof	2	May 2005	1	Jahnke et al. (2007b)
<0.1 <0.3	11.7 <0.1	34.4 <0.27	17.2 <0.40		g p	Kjeller, Norway	1	Nov, Dec 2005	1	Barber et al. (2007)
1.4 <2.7	4.95 <0.2	11.3 <1.9	7.8 <0.5	<25 <6.3	g p	Mace Head, Ireland	1	Mar 2006	1	Barber et al. (2007)
0.1	23	50	21	16	g	Geesthacht, Germany	121	Apr 2007– Jun 2008	1	Dreyer et al. (2009a)
0.3	22	62	21	13	g	Barsbüttel, Germany	117	Apr 2007– Jun 2008	1	Dreyer et al. (2009a)
Remote										
	(<0.4–5.0)				g	Okinawa, Japan	18	Mar–May 2004	1	Piekarz et al. (2007)

Table 13 (continued)

Concentration (pg/m^3)

$x = 3$ (4:2)	$x = 5$ (6:2)	$x = 7$ (8:2)	$x = 9$ (10:2)	$x = 11$ (12:2)	Phase	Location	n	Date collected	Method	References
	(<0.4–4.0)	(<0.9–18)	(<1–7.5)		g	Mount Bachelor, Oregon	8	Apr–Jul 2004	1	Piekarz et al. (2007)
	(<0.4–2.1)	(<0.9–2.0)	(<1–2.8)		p	Mount Bachelor, Oregon	8	Apr–Jul 2004	1	Piekarz et al. (2007)
	(nd–20)	(nd–19)	(nd–4.1)		g+p	Canadian High Arctic	10	Jul, Aug 2004	1	Stock et al. (2007)
	2.65 (<1.1–5.98) <0.001	11.4 (4.16–22.7) 3.50 (1.07–8.37)	6.27 (1.45–16.4) 0.80 (0.29–1.57)		g p	North Atlantic/ Canadian Arctic	20	Jul 2005	1	Shoeib et al. (2006)
	157 (140,174)	176 (163,190)	46 (44,48)		g+p	North Sea	1	Oct 2005	1	Jahnke et al. (2007c)
	11 (8.4,14)	29 (21,36)	9.2 (6.5,12)		g+p	Atlantic Ocean (46 N/6 W – 45 N/4 W)	1	Oct 2005	1	Jahnke et al. (2007c)

Table 13 (continued)

Concentration (pg/m³)										
$x=3$ (4:2)	$x=5$ (6:2)	$x=7$ (8:2)	$x=9$ (10:2)	$x=11$ (12:2)	Phase	Location	n	Date collected	Method	References
	9.4	15	3.3		g+p	Atlantic Ocean (40 N/10 W–18 N/20 W)	1	Oct 2005	1	Jahnke et al. (2007c)
	14 (10,19)	18 (17,20)	4.8 (4.7,4.9)		g+p	Atlantic Ocean (30 N/16 W–18 N/20 W)	1	Oct 2005	1	Jahnke et al. (2007c)
	20 (19,21)	42 (35,48)	7.7 (7.1,8.3)		g+p	Atlantic Ocean (18 N/20 W–6 N/16 W)	1	Oct, Nov 2005	1	Jahnke et al. (2007c)
	nd	13 (12,14)	2.9 (2.8,2.9)		g+p	Atlantic Ocean (6 N/16 W–4 S/8 W)	1	Nov 2005	1	Jahnke et al. (2007c)
	nd	3.1 (3.0,3.2)	1.2 (1.0, 1.4)		g+p	Atlantic Ocean (8 S/5 W–17 S/2 E)	1	Nov 2005	1	Jahnke et al. (2007c)
	nd	2.4 (2.0,2.8)	0.9 (0.8,1.0)		g+p	Atlantic Ocean (17 S/2 E–26 S/9 E)	1	Nov 2005	1	Jahnke et al. (2007c)

Table 13 (continued)

Concentration (pg/m³)					Phase	Location	n	Date collected	Method	References
x = 3 (4:2)	x = 5 (6:2)	x = 7 (8:2)	x = 9 (10:2)	x = 11 (12:2)						
(<0.4–16)	(<0.4–16)	(<0.9–44)	(<1–42)		g	Mount Bachelor, Oregon	34	Apr, May 2006	1	Piekarz et al. (2007)
	(<0.4–<1.2)	(<0.9–27)	(<1–26)		p					
(nd–1.2)	(1.7–165)	(10–124)	(1.6–53)	(0.6–35)	g	North Atlantic	23	Apr, May 2007	1	Dreyer et al. (2009b)
0.0	0.1	0.4	0.1	0.4	p	Atlantic, Southern Oceans and Baltic Sea	99	Apr 2007– May 2009	1	Dreyer et al. (2009b)
1.4	5.0	11	2.8		g	Elbe River, Germany	2	Oct, Nov 2007	1	Dreyer and Ebinghaus (2009)
nd	nq	nd	nd		p					
nd	8.6	56	15		g	German Bight	2	Oct, Nov 2007	1	Dreyer and Ebinghaus (2009)
nd	nq	nd	nd		p					
nd	5.8	13	3.5		g	German Bight	2	Oct, Nov 2007	1	Dreyer and Ebinghaus (2009)
nd	nq	nd	nd		p					
2.3	5.7	16	5.7		g	North Sea	2	Oct, Nov 2007	1	Dreyer and Ebinghaus (2009)
nd	nd	nd	nd		p					

Table 13 (continued)

Concentration (pg/m³)					Phase	Location	n	Date collected	Method	References
x = 3 (4:2)	x = 5 (6:2)	x = 7 (8:2)	x = 9 (10:2)	x = 11 (12:2)						
1.8	3.3	8.2	2.1		g	German Bight	2	Oct, Nov 2007	1	Dreyer and Ebinghaus (2009)
nd	nq	nd	nd		p					
1.4	15	28	6.7		g	German Bight	2	Oct, Nov 2007	1	Dreyer and Ebinghaus (2009)
nd	nq	nq	nd		p					
nd	(1.1–40)	(1.5–54)	(nd–90)	(1.5–165)	g	South Atlantic	27	Oct, Nov 2007	1	Dreyer et al. (2009b)
nd	(nd–34)	(7.1–77)	(4.3–45)	(0.3–20)	g	South Atlantic	13	Jan 2008	1	Dreyer et al. (2009b)
nd	(1.6–102)	(10–79)	(1.3–33)	(0.1–13)	g	Baltic Sea	34	Jun, Jul 2008	1	Dreyer et al. (2009b)
nd	(12–26)	(6.5–50)	(1.8–11)	(0.4–6.2)	g	North Atlantic	9	Aug 2008	1	Dreyer et al. (2009b)
nd	(nd–3.1)	(3.4–8.0)	(0.7–2.2)	(nd–0.6)	g	South Atlantic	7	Oct–Dec 2008	1	Dreyer et al. (2009b)
nd	(nd–0.9)	(1.8–11)	(1.1–4.2)	(nd–1.1)	g	Southern Ocean	14	Dec 2008–May 2009	1	Dreyer et al. (2009)

Mean concentration given where applicable, with ranges in brackets. Numbers appear as published in the original studies; nd = below limit of detection; nq = below limit of quantification; g = gas phase; p = particulate phase. Methods: (1) high-volume quartz fiber filter (QFF)/polyurethane foam (PUF)/XAD/PUF collection and extraction method originally described by Martin et al. (2002) and subsequent modifications thereof; (2) low-volume solid-phase extraction collection; (3) high-volume QFF/PUF/activated carbon fiber felt (ACF) collection; (4) passive ACF collection; (5) passive PUF collection
[a]Ranges estimated from bar graph
[b]Geometric mean given

Table 14 Summary of published atmospheric concentrations of FTOs ($CF_3(CF_2)_xCH=CH_2$)

Concentration (pg/m^3)									
x = 5 (6:2)	x = 7 (8:2)	x = 9 (10:2)	x = 11 (12:2)	Phase	Location	n	Date collected	Method	References
Urban									
		0.7		g	Manchester, UK	1	Feb, Mar 2005	1	Barber et al. (2007)
		0.3		p					
<2.6	24.7	3.1	3.3	g	Manchester, UK	1	Nov, Dec 2005	1	Barber et al. (2007)
<5.2	<0.8	<0.6	<0.8	p					
Semi-urban									
		0.2		g	Hazelrigg, UK	1	Feb, Mar 2005	1	Barber et al. (2007)
		<0.4		p					
2.7	11.9	2.0	5.2	g	Hazelrigg, UK	1	Nov 2005–Feb 2006	1	Barber et al. (2007)
<5.2	<0.8	<0.6	<0.8	p					
Rural									
		2.0		g	Kjeller, Norway	1	Nov, Dec 2005	1	Barber et al. (2007)
		<0.1		p					
<2.7	<0.4	1.5	2.2	g	Mace Head, Ireland	1	Mar 2006	1	Barber et al. (2007)
<5.4	<0.8	<0.6	<0.9	p					
Remote									
nd	nd	(<LOD–1.6)	(<LOD–2.2)	g	Hedo Station, Okinawa, Japan	18	Mar–May 2004	1	Piekarz et al. (2007)
nd	nd	nd	nd	p					

Mean concentration given where applicable, with ranges in brackets. Numbers appear as published in the original studies; nq = below limit of quantification; g = gas phase; p = particulate phase. Methods: (1) high-volume quartz fiber filter (QFF)/polyurethane foam (PUF)/polyurethane foam (PUF)/XAD/PUF collection and extraction method originally described by Martin et al. (2002) and subsequent modifications thereof; (2) low-volume solid-phase extraction collection; (3) high-volume QFF/PUF/activated carbon fiber felt (ACF) collection; (4) passive ACF collection; (5) passive PUF collection addition, the 8:2 FTO ($CF_3(CF_2)_xCH=CH_2$, $x = 7$) is the dominant one observed in the atmosphere, and concentrations of FTOs found are in the low pg/m^3 range.

FTAcs

Measurements of FTAcs have been attempted in a few studies performed since 2005. Results are shown in Table 15. Of the studies that included FTAcs, only a few produced detections. This is probably attributable to the short atmospheric lifetime of FTAcs. Concentrations tend to be less than 100 pg/m^3, although the levels of 8:2 FTAc ($CF_3(CF_2)_xCH_2CH_2OC(O)CH=CH_2$, $x = 7$) tend to be higher. Detections are too sparse to determine specific spatial trends; however, levels in Japan appear to be higher. In particular, a measurement from Higashiyodogawa, in Osaka City, Japan, shows exceptionally high levels of 8:2 FTAc (2,953 pg/m^3 (Oono et al. 2008b)). This result suggests the proximity of a point source and demonstrates the need for high spatial resolution for short-lived precursors, such as FTAcs.

4.5 Perfluorosulfonamides

4.5.1 Potential Sources to the Atmosphere

Polyfluorinated chemicals, used for their hydro- and lipophobic properties, are synthesized by either the telomerization or electrochemical process. Perfluorosulfonamides have been produced by the electrochemical process, in which products are characterized by the presence of a mixture of linear and branched carbon chains. This process was primarily used to produce perfluorosulfonyl fluoride ($CF_3(CF_2)_xSO_2F$). This is the synthetic precursor to NAFSAs, which can subsequently be converted to NAFSEs. Both NAFSAs and NAFSEs are typically converted into other products for commercial use, including phosphate esters for use in food packaging and polymers for fabric and carpet stain treatment (3M Company 1999). The incorporation of volatile compounds into polymers could lead to the presence of unreacted compounds in commercial products. Dinglasan-Panlilio and Mabury (2006) observed unreacted NMeFOSE in a product produced by electrochemical fluorination.

The presence of these unreacted materials, or residuals, in commercial products is a potential source of PFA precursors to the atmosphere. High levels of perfluorosulfonamides have been observed in indoor air (Barber et al. 2007; Jahnke et al. 2007d; Shoeib et al. 2004, 2005). Indoor concentrations were consistently much higher than those observed outdoors (Barber et al. 2007; Jahnke et al. 2007d; Shoeib et al. 2004, 2005).

Industry has responded to the presence of perfluorinated acids in wildlife. In 2001–2002, the largest producer of perfluorooctanesulfonyl fluoride, the precursor to PFOS and eight-carbon NAFSAs and NAFSEs, voluntarily removed the products from the market (3M Company 2000). These compounds have largely been

Table 15 Summary of published atmospheric concentrations of FTACs ($CF_3(CF_2)_xCH_2CH_2OC(O)CH=CH_2$)

Concentration (pg/m³)			Phase	Location	n	Date collected	Method	References
$x = 5$ (6:2)	$x = 7$ (8:2)	$x = 9$ (10:2)						
Urban								
nd			g+p	Hamburg	7	Apr, May 2005	1	Jahnke et al. (2007a)
nd			g+p	Hamburg	2	Apr 2005	1	Jahnke et al. (2007b)
	4[a] (nd–101)		g+p	Kyoto City, Japan	10	Oct–Dec 2006	1	Oono et al. (2008b)
	16[a] (22–74)		g+p	Osaka City, Japan	4	Sep–Dec 2006	3	Oono et al. (2008b)
	485[a] (nd–2953)		g+p	Osaka City, Japan	10	Oct, Nov 2006	3	Oono et al. (2008b)
	7.4[a] (<3–480)		g	Keihan Area, Japan	15	Apr–Jun 2007	4	Oono et al. (2008a)
	3.0[a] (<3–21.7)		g	Across Japan	18	Apr–Jun 2007	4	Oono et al. (2008a)
4.2	1.5	6.6	g	Hamburg	2	Oct, Nov 2007	1	Dreyer and Ebinghaus (2009)
nq	nq	nq	p					
Semi-urban								
0.2	(0.9,1.2)	(0.4,0.6)	g+p	Hamburg	2	Jun 2006	1	Dreyer et al. (2008)
(0.4–5.4)	(2.9–9.8)	(1.4–4.2)	g	Barsbuettel	3	Oct, Nov 2007	1	Dreyer and Ebinghaus (2009)

Table 15 (continued)

Concentration (pg/m³)			Phase	Location	n	Date collected	Method	References
$x=5$ (6:2)	$x=7$ (8:2)	$x=9$ (10:2)						
(1.0–7.2)	(1.7–14)	(0.7–4.6)	g	Hamburg	5	Oct, Nov 2007	1	Dreyer and Ebinghaus (2009)
Rural								
nd			g+p	Waldhof	2	May 2005	1	Jahnke et al. (2007b)
nd			g+p	Waldhof	4	May, Jun 2005	1	Jahnke et al. (2007a)
1.6	2.5	2.6	g	Geesthacht, Germany	121	Apr 2007–Jun 2008	1	Dreyer et al. (2009a)
1.9	4.2	2.5	g	Barsbüttel, Germany	117	Apr 2007–Jun 2008	1	Dreyer et al. (2009a)
Remote								
(nd–4.1)	(<0.7–5.9) (<0.7–4.3)		g p	Mount Bachelor, Oregon	34	Apr, May 2006	1	Piekarz et al. (2007)
(0.1–15)		(<0.1–4.5)	g	North Atlantic	23	Apr, May 2007	1	Dreyer et al. (2009b)
0.0	0.0	0.0	p	Atlantic, Southern Oceans and Baltic Sea	99	Apr 2007–May 2009	1	Dreyer et al. (2009b)

Table 15 (continued)

Concentration (pg/m³)				Phase	Location	n	Date collected	Method	References
$x = 5$ (6:2)	$x = 7$ (8:2)	$x = 9$ (10:2)							
nd	2.2	0.9	g	Elbe River, Germany	2	Oct, Nov 2007	1	Dreyer and Ebinghaus (2009)
nd	nd	nd	p						
5.7	3.2	1.3	g	German Bight	2	Oct, Nov 2007	1	Dreyer and Ebinghaus (2009)
nd	nd	nd	p						
nd	1.9	0.8	g	German Bight	2	Oct, Nov 2007	1	Dreyer and Ebinghaus (2009)
nq	nd	nd	p						
nd	1.7	1.2	g	North Sea	2	Oct, Nov 2007	1	Dreyer and Ebinghaus (2009)
nd	nd	nd	p						
nq	1.8	0.8	g	German Bight	2	Oct, Nov 2007	1	Dreyer and Ebinghaus (2009)
nd	nd	nd	p						
nd	3.6	1.2	g	German Bight	2	Oct, Nov 2007	1	Dreyer and Ebinghaus (2009)
nd	nq	nd	p						
(nd–16)	(nd–3.5)	(nd–1.6)	g	South Atlantic	27	Oct, Nov 2007	1	Dreyer et al. (2009b)

Table 15 (continued)

Concentration (pg/m^3)			Phase	Location	n	Date collected	Method	References
$x = 5$ (6:2)	$x = 7$ (8:2)	$x = 9$ (10:2)						
(nd–5.5)	(nd–3.6)	nd	g	South Atlantic	13	Jan 2008	1	Dreyer et al. (2009b)
(nd–7.3)	(nd–3.7)	(nd–7.0)	g	Baltic Sea	34	Jun, Jul 2008	1	Dreyer et al. (2009b)
(nd–8.9)	(nd–5.2)	(nd–0.3)	g	North Atlantic	9	Aug 2008	1	Dreyer et al. (2009b)
nd	nd	nd	g	South Atlantic	7	Oct–Dec 2008	1	Dreyer et al. (2009b)
(nd–0.9)	(nd–0.2)	nd	g	Southern Ocean	14	Dec 2008–May 2009	1	Dreyer et al. (2009b)

Mean concentration given where applicable, with ranges in brackets. Numbers appear as published in the original studies; nd = below limit of detection; nq = below limit of quantification; g = gas phase; p = particulate phase. Methods: (1) high-volume quartz fiber filter (QFF)/polyurethane foam (PUF)/XAD/PUF collection and extraction method originally described by Martin et al. (2002) and subsequent modifications thereof; (2) low-volume solid-phase extraction collection; (3) high-volume QFF/PUF/activated carbon fiber felt (ACF) collection; (4) passive ACF collection; (5) passive PUF collection

[a]Geometric mean given

replaced with their four-carbon equivalents, which as a result, should be observed in progressively higher concentrations in the environment.

4.5.2 Atmospheric Concentrations

NAFSA

The NAFSAs have been a subject of several monitoring studies from various locations (Table 16), typically alongside FTOHs. However, as with the FTOHs, studies have been concentrated in North America and Europe. NAFSAs with a perfluorinated chain of eight carbons (*N*-alkyl-perfluorooctanesulfonamides (NAFOSAs)) were first measured in 2001 by Martin et al. (2002) and have been detected many times subsequently, while those with a perfluorinated chain of four (*N*-alkyl-perfluorobutanesulfonamides (NAFBSAs)) have been included since 2005. NAFSAs are found at low concentrations, typically in the low pg/m^3 range. Levels of NAFOSAs and NAFBSAs in the atmosphere are comparable, despite the differences in production trends. NAFSAs appear to partition between the gas and particulate phase, though NAFOSAs appear more frequently in the particulate phase. The low levels detected in the atmosphere make it difficult to discern trends; nevertheless, it is clear that concentrations in remote areas are lower than those of other areas.

NAFSE

The NAFSEs have typically been measured alongside NAFSAs in studies conducted primarily in North America and Europe (Table 17). NAFSEs, with a perfluorinated chain of eight carbons (*N*-alkyl-perfluorooctanesulfamido ethanols (NAFOSEs)), have been the focus of studies since 2001, while those containing a four-carbon perfluorinated chain have been incorporated since 2005 (*N*-alkyl-perfluorobutanesulfamido ethanols (NAFBSEs)). Spatial variability is evident for NAFOSEs, with levels in urban areas typically higher than those in remote areas.

5 Impact of Precursors on Environmental Perfluorinated Acid (PFA) Levels

5.1 Trifluoroacetic Acid (TFA)

Atmospheric TFA is primarily formed through perfluoroacyl halide hydrolysis, and presumably this occurs in cloud particles. TFA is very soluble and likely to stay in solution and to rain-out, on a similar timescale to nitric acid (approximately 9 d) (Kotamarthi et al. 1998). However, it is possible that TFA could be liberated to the gas phase where it can react with hydroxyl radicals on a timescale of 100–230 d (Carr et al. 1994; Hurley et al. 2004c). This process has been estimated to reduce the amount of TFA present in rainwater by less than 5% (Kanakidou et al. 1995). As a

Table 16 Summary of published atmospheric concentrations of NAFSAs (CF₃(CF₂)ₓSO₃N(R)H)

Mean concentration in pg/m³ (range)

PFOSA	NMeFOSA	NEtFOSA	NMeFBSA	Phase	Location	n	Date	Method	References
Urban									
		14		g+p	Toronto, ON	2	Feb–Apr 2001	1	Martin et al. (2002)
		nd		g+p	Toronto, ON	3	Nov 2001	1	Stock et al. (2004)
		(20–130)[a]		g+p	Reno, NV	3	Nov 2001	1	Stock et al. (2004)
		nd		g+p	Winnipeg, MB	3	Nov 2001	1	Stock et al. (2004)
		nd		g+p	Ottawa, ON	7	Winter 2002/2003	5	Shoeib et al. (2005)
nd	7 (6.8,7.2)	3 (2.5,2.6)		g+p	Hamburg	2	Apr 2005	1	Jahnke et al. (2007b)
	6.1	9.6		g	Manchester, UK	1	Feb, Mar 2005	1	Barber et al. (2007)
<1.6	1.5	0.7		p					
	9.0 (3.4–20)	3.1 (1.3–5.9)	0.4	g	Hamburg	7	Apr, May 2005	1	Jahnke et al. (2007a)
	<0.2	nd	<0.2	p					
	<5.3	<3.7		g	Manchester, UK	1	Nov, Dec 2005	1	Barber et al. (2007)
<0.2	<9.1	<6.5		p					
	0.9 (nd–2.5)	0.3 (nd–0.9)		g+p	Hamburg	5	Apr 2006	2	Jahnke et al. (2007d)
	1.7 (1–2.6)	0.8 (0.5–1.1)		g+p	Hamburg	4	Apr 2006	1	Jahnke et al. (2007d)
7.3	2.4	0.5	3.4	g	Hamburg	2	Oct, Nov 2007	1	Dreyer and Ebinghaus (2009)
nd	3.4	2.4	nd	p					
Semi-urban									
		(nd–20)[a]		g+p	Griffin, GA	5	Nov 2001	1	Stock et al. (2004)
	5.5	(1–125)[a]		g+p	Cleves, OH	3	Nov 2001	1	Stock et al. (2004)
	<1.1	7.9		g	Hazelrigg, UK	1	Feb, Mar 2005	1	Barber et al. (2007)
<2.1		0.4		p					
	<5.3	<3.7	0.2	g	Hazelrigg, UK	1	Nov 2005–Feb 2006	1	Barber et al. (2007)
0.2	<9.1	<6.5	<0.1	p					

Table 16 (continued)

Mean concentration in pg/m^3 (range)

PFOSA	NMeFOSA	NEtFOSA	NMeFBSA	Phase	Location	n	Date	Method	References
(0.4,0.8)	(5.4,6.9)	(2.0,2.3)	(0.4,0.8)	g+p	Hamburg	2	Jun 2006	1	Dreyer et al. (2008)
(nd–1.6)	(1.3–2.1)	(1.4–2.7)	(1.4–4.9)	g	Barsbuettel	3	Oct, Nov 2007	1	Dreyer and Ebinghaus (2009)
(nd–1.4)	(1.0–4.8)	(0.7–7.1)	(1.3–10)	g	Hamburg	5	Oct, Nov 2007	1	Dreyer and Ebinghaus (2009)
Rural									
nd	8 (5.1,10)	(1–16)[a]		g+p	Long Point, ON	3	Nov 2001	1	Stock et al. (2004)
	7.0 (3.8–11)	3 (2.6,3.2)		g+p	Waldhof	2	May 2005	1	Jahnke et al. (2007b)
	(nd–<0.2)	2.6 (1.5–3.4)		g	Waldhof	4	May, Jun 2005	1	Jahnke et al. (2007a)
		nd		p					
0.78	5.5	5		g	Kjeller, Norway	1	Nov, Dec 2005	1	Barber et al. (2007)
	<0.43	<0.71		p					
<0.56	<4.9	<1.6	<0.1	g	Mace Head, Ireland	1	Mar 2006	1	Barber et al. (2007)
	<4.7	<2.9	<0.2	p					
0.8	2.9	1.5	3.6	g	Geesthacht, Germany	121	Apr 2007–Jun 2008	1	Dreyer et al. (2009a)
1.0	2.6	1.3	3.0	g	Barsbuttel, Germany	117	Apr 2007–Jun 2008	1	Dreyer et al. (2009a)
Remote									
		nd–2.2		g	Lake Ontario	8	Aug 2003		Boulanger et al. (2005)
	(<0.4–1.2)	(<0.4–1.2)		g	Mount Bachelor, Oregon	8	Apr–Jul 2004	1	Piekarz et al. (2007)
				p					

Table 16 (continued)

Mean concentration in pg/m³ (range)

PFOSA	NMeFOSA	NEtFOSA	NMeFBSA	Phase	Location	n	Date	Method	References
	(nd–64)	(nd–19)	(nd–19)		Canadian High Arctic	10	Jul, Aug 2004	1	Stock et al. (2007)
	3.8 (3.4,4.2)	2.0 (1.7,2.2)		g+p	North Sea	1	Oct 2005	1	Jahnke et al. 2007c
	1.9 (1.7,2.0)	0.8 (0.7,0.8)		g+p	Atlantic Ocean (46 N/6 W–45 N/4 W)	1	Oct 2005	1	Jahnke et al. (2007c)
	1.1	<0.3		g+p	Atlantic Ocean (40 N/10 W–18 N/20 W)	1	Oct 2005	1	Jahnke et al. (2007c)
	1.6 (1.6,1.6)	0.7 (0.7,0.7)		g+p	Atlantic Ocean (30 N/16 W–18 N/20 W)	1	Oct 2005	1	Jahnke et al. (2007c)
	2.3 (2.1,2.4)	1.1 (0.9,1.3)		g+p	Atlantic Ocean (18 N/20 W–6 N/16 W)	1	Oct, Nov 2005	1	Jahnke et al. (2007c)
	0.6 (0.5,0.6)	(<0.3, nd)		g+p	Atlantic Ocean (6 N/16 W–4 S/8 W)	1	Nov 2005	1	Jahnke et al. (2007c)
	1.2 (0.4,2.0)	(<0.3, 1.3)		g+p	Atlantic Ocean (8 S/5 W–17 S/2 E)	1	Nov 2005	1	Jahnke et al. (2007c)
	0.5 (0.4,0.5)	nd		g+p	Atlantic Ocean (17 S/2 E–26 S/9 E)	1	Nov 2005	1	Jahnke et al. (2007c)
		(<0.4–1.9)		g	Mount Bachelor, Oregon	34	Apr, May 2006	1	Piekarz et al. (2007)
		(<0.4–1.9)		p					
(nd–7.4)	(0.4–7.9)	(<0.1–67)	(<0.3–5.6)	g	North Atlantic	23	Apr, May 2007	1	Dreyer et al. (2009b)

Table 16 (continued)

Mean concentration in pg/m³ (range)

PFOSA	NMeFOSA	NEtFOSA	NMeFBSA	Phase	Location	n	Date	Method	References
0.0	0.3	0.2	0.0	p	Atlantic, Southern Oceans and Baltic Sea	99	Apr 2007–May 2009	1	Dreyer et al. (2009b)
2.5	3.7	1.7	7.1	g	Elbe River, Germany	2	Oct, Nov 2007	1	Dreyer and Ebinghaus (2009)
nd	4.2	3.1	nd	p					
1.9	1.5	0.8	5.8	g	German Bight	2	Oct, Nov 2007	1	Dreyer and Ebinghaus (2009)
nd	3.3	2.1	nd	p					
3.4	2.5	0.4	3.1	g	German Bight	2	Oct–Nov 2007	1	Dreyer and Ebinghaus (2009)
nd	nd	nd	nd	p					
2.8	3.3	1.4	3.4	g	North Sea	2	Oct, Nov 2007	1	Dreyer and Ebinghaus (2009)
nd	nd	nd	nd	p					
nq	3.9	1.5	6.0	g	German Bight	2	Oct, Nov 2007	1	Dreyer and Ebinghaus (2009)
nd	9.2	7.9	nd	p					
nd	3.1	1.1	4.7	g	German Bight	2	Oct, Nov 2007	1	Dreyer and Ebinghaus (2009)
nd	5.6	3.1	nd	p					
nd	(nd–5.5)	(0.5–4.4)	(nd–4.6)	g	South Atlantic	27	Oct, Nov 2007	1	Dreyer et al. (2009b)
nd	(nd–5.5)	(nd–4.9)	(nd–5.6)	g	South Atlantic	13	Jan 2008	1	Dreyer et al. (2009b)
nd	(0.3–14)	(nd–3.7)	(nd–4.4)	g	Baltic Sea	34	Jun, Jul 2008	1	Dreyer et al. (2009b)
(nd–5.2)	(0.4–7.7)	(<0.1–3.1)	(0.3–6.0)	g	North Atlantic	9	Aug 2008	1	Dreyer et al. (2009b)
nd	(1.5–3.9)	nd	(nd–4.5)	g	South Atlantic	7	Oct–Dec 2008	1	Dreyer et al. (2009b)
nd	(nd–2.3)	(nd–0.1)	(nd–1.3)	g	Southern Ocean	14	Dec 2008–May 2009	1	Dreyer et al. (2009b)

Mean concentration given where applicable, with ranges in brackets. Numbers appear as published in the original studies; nd = below limit of detection; nq = below limit of quantification; g = gas phase; p = particulate phase. Methods: (1) high-volume quartz fiber filter (QFF)/polyurethane foam (PUF)/XAD/PUF collection and extraction method originally described by Martin et al. (2002) and subsequent modifications thereof; (2) low-volume solid-phase extraction collection; (3) high-volume QFF/PUF/activated carbon fiber felt (ACF) collection; (4) passive ACF collection; (5) passive PUF collection

[a]Ranges estimated from bar graph

Table 17 Summary of published atmospheric concentrations of NAFSEs $(CF_3(CF_2)_xSO_3N(R)CH_2CH_2OH)$

| Mean concentration in pg/m³ (range) | | | | | | | | | |
NMeFOSE	NEtFOSE	NMeFBSE	NEtFBSE	Phase	Location	n	Date	Method	References
Urban									
101(86–123)[a]	205 (51–393)			g+p	Toronto, ON	4	Feb–Apr 2001	1	Martin et al. (2002)
(30–39)[a]	(nd–173)[a]			g+p	Toronto, ON	3	Nov 2001	1	Stock et al. (2004)
359 (20–35)[a]	199 (80–340)[a]			g+p	Reno, NV	3	Nov 2001	1	Stock et al. (2004)
(12–30)[a]	nd			g+p	Winnipeg, MB	3	Nov 2001	1	Stock et al. (2004)
83	88			g+p	Ottawa, ON	7	Winter 2002/2003	5	Shoeib et al. (2005)
23.7	6.4			g	Manchester, UK	1	Feb, Mar 2005	1	Barber et al. (2007)
24	11			p					
30 (30,29)	10 (9.3,11)			g+p	Hamburg	2	Apr 2005	1	Jahnke et al. (2007b)
30 (3.0–89)	7.6 (0.5–27)			g	Hamburg	7	Apr, May 2005	1	Jahnke et al. (2007a)
11 (2.3–18)	6.7 (2.4–12)			p					
<53.5	<2.2	<2.5		g	Manchester, UK	1	Nov, Dec 2005	1	Barber et al. (2007)
<1.8	<4.3	<4.9		p					
8.0	2.33			g	Toronto, ON	3	Mar 2006	1	Shoeib et al. (2006)
(5.38–11.8)	(1.04–3.01)			p					
4.20	0.96								
(2.67–6.51)	(0.40–1.68)								
5.8 (nd–9.3)	nd			g+p	Hamburg	5	Apr 2006	2	Jahnke et al. (2007d)
10 (6.9–14)	2.6 (1.4–4.1)			g+p	Hamburg	4	Apr 2006	1	Jahnke et al. (2007d)
1.4	nd	2.5		g	Hamburg	2	Oct, Nov 2007	1	Dreyer and Ebinghaus (2009)
9.0	6.9	nd		p					
Semi-urban									
(16.0,31.7)	(8.47,9.79)			g	Ontario, Canada	2	Nov 2000, Mar 2001	1	Shoeib et al. (2004)
(30–>600)[a]	(nd–80)[a]			g+p	Griffin, GA	5	Nov 2001	1	Stock et al. (2004)
(2–58)[a]	nd			g+p	Cleves, OH	3	Nov 2001	1	Stock et al. (2004)

Table 17 (continued)

Mean concentration in pg/m³ (range)

NMeFOSE	NEtFOSE	NMeFBSE	NEtFBSE	Phase	Location	n	Date	Method	References
	(nd–1.0)			g	Lake Ontario	8	Aug 2003		Boulanger et al. (2005)
24	9.2			g	Hazelrigg, UK	1	Feb, Mar 2005	1	Barber et al. (2007)
12.1	6.9			p					
<53.5	65.7	<2.5		g	Hazelrigg, UK	1	Nov 2005–Feb 2006	1	Barber et al. (2007)
<1.8	6.9	<4.9		p					
(1.9,2.3)	(0.4,0.6)	(2.6,2.9)		g+p	Hamburg	2	June 2006	1	Dreyer et al. (2008)
(2.3–3.8)	(0.4–0.8)	(2.9–6.4)		g	Barsbuettel	3	Oct, Nov 2007	1	Dreyer and Ebinghaus (2009)
(0.6–7.4)	(0.3–1.9)	(1.0–5.5)		g	Hamburg	5	Oct, Nov 2007	1	Dreyer and Ebinghaus (2009)
Rural									
35 (34,36)	76 (68,85)			g+p	Long Point, ON	2	Feb–Apr 2001	1	Martin et al. (2002)
(18–36)[a]	(10–17)[a]			g+p	Long Point, ON	3	Nov 2001	1	Stock et al. (2004)
6.5 (0.5–11)	11 (<0.3–23)			g	Waldhof	4	May, Jun 2005	1	Jahnke et al. (2007a)
2.4 (0.9–4.7)	5.9 (1.2–15)			p					
10 (8.9,11)	23 (21,24)			g+p	Waldhof	2	May 2005	1	Jahnke et al. (2007b)
48.9	29.5			g	Kjeller, Norway	1	Nov, Dec 2005	1	Barber et al. (2007)
3.6	3.4			p					
<79.6	<52.4	<14.5		g	Mace Head, Ireland	1	Mar 2006	1	Barber et al. (2007)
<18.9	<7.5	<20.0		p					
2.0	1.0	1.7		g	Geesthact, Germany	121	Apr 2007–Jun 2008	1	Dreyer et al. (2009a)
2.2	1.0	2.7		g	Barsbüttel, Germany	117	Apr 2007–Jun 2008	1	Dreyer et al. (2009a)
Remote									
(<1–25)	(<1–8.7)			g	Mount Bachelor, Oregon	18	Mar–May 2004	1	Piekarz et al. (2007)
(<1–21)	(<1–6.9)			p					

Table 17 (continued)

Mean concentration in pg/m³ (range)				Phase	Location	n	Date	Method	References
NMeFOSE	NEtFOSE	NMeFBSE	NEtFBSE						
(<1–4.8)	(<1–7.1)			g	Mount Bachelor, Oregon	8	Apr–Jul 2004	1	Piekarz et al. (2007)
(<1–3.9)	(<1–4.5)			p					
(nd–39)	(nd–28)	(nd–34)	nd	g+p	Canadian High Arctic	10	Jul, Aug 2004	1	Stock et al. (2007)
8.30 (<1.9–23.6)	1.87 (<1.0–5.17)			g	North Atlantic/Canadian Arctic	20	Jul 2005	1	Shoeib et al. (2006)
3.53 (<1.7–15.0)	1.05 (<0.001–5.5)			p					
19 (16,22)	8.9 (5.9,11.8)			g+p	North Sea	1	Oct 2005	1	Jahnke et al. (2007c)
3.9 (3.0,4.8)	nd			g+p	Atlantic Ocean (46 N/6 W– 45 N/4 W)	1	Oct 2005	1	Jahnke et al. (2007c)
1.6	0.9			g+p	Atlantic Ocean (40 N/10 W– 18 N/20 W)	1	Oct 2005	1	Jahnke et al. (2007c)
3.1 (2.9,3.3)	nd			g+p	Atlantic Ocean (30 N/16 W– 18 N/20 W)	1	Oct 2005	1	Jahnke et al. (2007c)
7.4 (7.3,7.5)	1.5 (1.2,1.7)			g+p	Atlantic Ocean (18 N/20 W– 6 N/16 W)	1	Oct, Nov 2005	1	Jahnke et al. (2007c)
(nd, 0.9)	nd			g+p	Atlantic Ocean (6 N/16 W– 4 S/8 W)	1	Nov 2005	1	Jahnke et al. (2007c)
nd	nd			g+p	Atlantic Ocean (8 S/5 W– 17 S/2 E)	1	Nov 2005	1	Jahnke et al. (2007c)

Table 17 (continued)

Mean concentration in pg/m³ (range)									
NMeFOSE	NEtFOSE	NMeFBSE	NEtFBSE	Phase	Location	n	Date	Method	References
nd	nd			g+p	Atlantic Ocean (17 S/2 E–26 S/9 E)	1	Nov 2005	1	Jahnke et al. (2007c)
(<1–11)	(<1–<3)			g	Mount Bachelor, Oregon	34	Apr, May 2006	1	Piekarz et al. (2007)
(<1–9.3)	(<1–3.7)			p					
(<0.4–10)	(nd–5.2)	(<0.1–9.3)		g	North Atlantic	23	Apr, May 2007	1	Dreyer et al. (2009b)
0.4	0.4	0.1		p	Atlantic, Southern Oceans and Baltic Sea	99	Apr 2007–May 2009	1	Dreyer et al. (2009b)
1.3	nd	0.6		g	Elbe River, Germany	2	Oct, Nov 2007	1	Dreyer and Ebinghaus (2009)
nd	nd	nd		p					
0.9	0.1	1.4		g	German Bight	2	Oct, Nov 2007	1	Dreyer and Ebinghaus (2009)
7.0	5.5	nd		p					
2.0	nd	1.0		g	German Bight	2	Oct, Nov 2007	1	Dreyer and Ebinghaus (2009)
nd	4.4	nd		p					
2.5	0.3	1.3		g	North Sea	2	Oct, Nov 2007	1	Dreyer and Ebinghaus (2009)
nd	nd	nd		p					
2.2	0.3	0.6		g	German Bight	2	Oct, Nov 2007	1	Dreyer and Ebinghaus (2009)
13	15	nd		p					
1.2	nd	1.0		g	German Bight	2	Oct, Nov 2007	1	Dreyer and Ebinghaus (2009)
nd	8.0	nd		p					

Table 17 (continued)

Mean concentration in pg/m³ (range)				Phase	Location	n	Date	Method	References
NMeFOSE	NEtFOSE	NMeFBSE	NEtFBSE						
(0.9–3.7)	(nd–2.1)	(nd–2.4)		g	South Atlantic	27	Oct, Nov 2007	1	Dreyer et al. (2009b)
(nd–20)	(nd–7.9)	(1.0–209)		g	South Atlantic	13	Jan 2008	1	Dreyer et al. (2009b)
(0.1–10.5)	(nd–1.2)	(nd–4.1)		g	Baltic Sea	34	Jun, Jul 2008	1	Dreyer et al. (2009b)
(<0.4–2.5)	(<0.1–2.5)	(<0.1–3.4)		g	North Atlantic	9	Aug 2008	1	Dreyer et al. (2009b)
(0.3–1.3)	(nd–0.4)	(0.6–2.5)		g	South Atlantic	7	Oct–Dec 2008	1	Dreyer et al. (2009b)
(nd–4.5)	(nd–1.1)	(nd–0.7)		g	Southern Ocean	14	Dec 2008–May 2009	1	Dreyer et al. (2009b)

Mean concentration given where applicable, with ranges in brackets. Numbers appear as published in the original studies; nd = below limit of detection; nq = below limit of quantification; g = gas phase; p = particulate phase. Methods: (1) high-volume quartz fiber filter (QFF)/polyurethane foam (PUF)/XAD/PUF collection and extraction method originally described by Martin et al. (2002) and subsequent modifications thereof; (2) low-volume solid-phase extraction collection; (3) high-volume QFF/PUF/activated carbon fiber felt (ACF) collection; (4) passive ACF collection; (5) passive PUF collection

[a]Ranges estimated from bar graph

result, almost all TFA formed from precursors in the atmosphere will be rained-out into the aqueous environment.

TFA is present ubiquitously in the aqueous environment, including in deep ocean water (Scott et al. 2005) and in precipitation (Scott et al. 2006). The compound has been observed to be slightly phytoaccumulative to higher plants (Thompson et al. 1994, 1995) and phytotoxic to some species at low concentrations (Smyth et al. 1994). Although TFA itself was produced as a commercial product, production levels are relatively small, on the order of 1,000 t/year (Boutonnet et al. 1999) A summary of estimated direct and indirect contributions to TFA is shown in Table 18. The impact of atmospheric sources on environmental TFA levels has been evaluated in several studies. In one study, it was calculated that under very specific conditions, TFA derived from CFC-replacement compounds could accumulate in wetlands to levels that have been observed to be toxic (Tromp et al. 1995). However, Boutonnet et al. (1999) suggested that these conditions were unlikely to ever occur in the environment and such accumulation was improbable. The study of Tromp et al. (1995) only took into account degradation of HFCs and HCFCs. Other studies that have included examinations of TFA levels derived from saturated HFCs, HCFCs, and anesthetics were unable to account for observed environmental levels (Jordan and Frank 1999; Kotamarthi et al. 1998; Tang et al. 1998). In addition, Jordan and Frank (1999) observed that TFA levels were much higher in rivers in industrialized regions than in remote rivers. This suggests the potential contributions of shorter-lived precursors, such as HFOs and others shown in Table 19. The thermolysis of fluoropolymers during use of commercial products (Ellis et al. 2001) or incineration of those products (Jordan and Frank 1999) may also be adding to the environmental burden. More information regarding source levels and atmospheric concentrations is required to better understand the true impact of atmospheric oxidation of volatile chemicals on environmental levels of TFA.

Table 18 Estimated indirect and direct sources of trifluoroacetic acid (TFA)

		Chemical	Estimated TFA production (t/year)	References
Indirect sources	Halothane	$CF_3CHClBr$	520	Tang et al. (1998)
	Isoflurane	$CF_3CHClOCHF_2$	280	Tang et al. (1998)
	HCFC-123	CF_3CHCl_2	266	Velders et al. (2005)
	HCFC-124	CF_3CHFCl	4,440	Velders et al. (2005)
	HFC-134a	CF_3CH_2F	4,560	Velders et al. (2005)
	Pyrolysis of fluoropolymers		200	Jordan and Frank (1999)
		Total	10,266	
Direct sources	Direct production		~1,000	Boutonnet et al. (1999)
		Total	1,000	

Table 19 Perfluorinated acids formed from volatile precursors, as described in Sections 3.2 and 3.3

Commercial compound	PFCAs														PFSAs	
	C2 TFA	C3 PFPrA	C4 PFBA	C5 PFPeA	C6 PFHxA	C7 PFHpA	C8 PFOA	C9 PFNA	C10 PFDA	C11 PFUnA	C12 PFDoA	C13 PFTrA	C14 PFTeA	C15 PFPA	C4 PFBS	C8 PFOS
Halothane	✓															
Isofluorane	✓															
HCFC-124	✓															
HCFC-123	✓															
HCFC-225a		✓														
HFC-134a	✓															
HFC-125	✓															
HFC-329	✓	✓														
HFC-227	✓															
CF$_3$CF=CH$_2$	✓		✓													
CF$_3$CF=CHF	✓															
CF$_3$CF=CF$_2$	✓															
CF$_3$CF$_2$CF=CF$_2$		✓														
CF$_3$CF=CFCF$_3$	✓															
Fluorotelomer compounds																
4:2	✓	✓	✓	✓												
6:2	✓	✓	✓	✓	✓	✓										
8:2	✓	✓	✓	✓	✓	✓	✓	✓								
10:2	✓	✓	✓	✓	✓	✓	✓	✓	✓	✓						
12:2	✓	✓	✓	✓	✓	✓	✓	✓	✓	✓	✓	✓				
14:2	✓	✓	✓	✓	✓	✓	✓	✓	✓	✓	✓	✓	✓	✓		
NAFSA																
NEtFBSA	✓	✓	✓	✓	✓	✓	✓									
NEtFOSA	✓	✓	✓													
NAFSE																
NMeFBSE	✓	✓	✓	✓	✓	✓	✓								✓	
NMeFOSE	✓	✓	✓													✓

5.2 Perfluorooctanesulfonic Acid (PFOS), Perfluorooctanoic Acid (PFOA) and Perfluorononanoic Acid (PFNA)

A great deal of attention has been paid to the impact of atmospheric formation of PFOS, PFOA, and PFNA. All three of these compounds can be formed not only indirectly through atmospheric oxidation of volatile precursors, but have also been intentionally produced directly. As such, detailed studies are required to attribute the importance of direct and indirect sources.

Relative significance of direct and indirect sources depends on the proximity to either a source of directly emitted compound or to a source of precursor compounds. Potential mechanisms of transport are expected to be very different for PFOS, PFOA, and PFNA compared to their respective precursors. The low pK_as of PFOS, PFOA, and PFNA cause them to be ionized under most environmental conditions and to have low volatility and high water solubility. Thus, the primary repository for these compounds is the aqueous environment, which has been substantiated by ubiquitous observations of these compounds in ocean water (Yamashita et al. 2005). Movement of compounds via water is very slow compared to air, reducing the speed of long-range transport. Emissions data from a PFOA producer indicate that a small amount (5%) is emitted to air (Prevedouros et al. 2006). It has often been assumed that PFOA cannot undergo long-range atmospheric transport because of its low volatility and relatively high water solubility. This was made under the assumption that the pK_a of PFOA was low and that it would be present exclusively in the anionic form in the environment. Further recent studies have suggested that the pK_a could range from 0 (Goss 2008) to 3.8 (Burns et al. 2008), while it was observed in another recent study that PFOA behaved similarly to the highly acidic PFOS and has a pK_a <1 (Cheng et al. 2009). The results from certain experiments have suggested that PFOA could be liberated into the gas phase by the production of aerosols and subsequent aerosol–gas partitioning, thus facilitating long-range transport through the atmosphere of the directly emitted chemical (McMurdo et al. 2008). In a subsequent modeling study, this problem was examined and it was determined that long-range atmospheric transport of PFOA would be insignificant unless the pK_a was at least 3.5 (Armitage et al. 2009a). If, indeed, directly emitted PFOA was being transported through the air, it would be expected that PFNA might be transported via a similar mechanism and that precipitation levels would reflect their respective direct production levels. Despite the fact that PFOA is produced directly at levels of at least one order of magnitude higher than PFNA, PFOA and PFNA concentrations have been observed to be similar in precipitation samples from mid-latitudes and high latitudes (Scott et al. 2006; Young et al. 2007). In addition, it has been suggested that the linear isomer of PFOA could be more readily liberated from the aerosol to the gas phase than branched PFOA isomers (McMurdo et al. 2008). If this were the case, the rain-out of isomer-enriched aerosols should lead to enrichment of branched isomers in precipitation that occurs near source regions relative to precipitation in remote areas (Ellis and Webster 2009). Isomer analysis from temperate precipitation and a remote arctic lake that receives input solely from the atmosphere demonstrated similar ratios of branched to linear PFOA isomers

(De Silva et al. 2009). This implies that the mechanism of transport does not discriminate among isomers, which is inconsistent with the hypothesis of aerosol-mediated transport of directly produced PFOA. Thus, we can suggest that the dominant mechanism of transport of PFOA and PFNA is the same as the highly acidic PFOS and occurs through water.

The yield of PFOS, PFOA, and PFNA from atmospheric oxidation of precursor chemicals is dependent on the ratio of NO_x to HO_2 and RO_2 compounds. Typically, NO_x levels are high in urban environments and low in remote environments. As a result, the highest yields of PFOS, PFOA, and PFNA from precursor compounds are likely to be in remote regions. However, atmospheric concentrations of precursor compounds are also higher at mid-latitudes that are close to production facilities and greater population density (see Section 4). Indirect production of PFOA at mid-latitudes is evident from the presence of isomers of PFOA as well as perfluoroheptanoic acid (PFHpA) and perfluorohexanoic acid (PFHxA) in rainwater (De Silva et al. 2009). Although branched PFHpA and PFHxA are not directly produced, volatile precursors to PFOA that are known to contain branched isomers, NAFSAs and NAFSEs, can degrade via the perfluorinated radical mechanism described in *Perfluorinated Radical Mechanism* Section to also form branched PFCAs of shorter-chain lengths.

The predicted deposition flux of PFOA and PFNA formed from 8:2 FTOH is highest in areas close to the sites of manufacture and use of the compounds, while the flux in remote regions is approximately one to two orders of magnitude lower (Yarwood et al. 2007). However, overall contamination in remote regions is much lower, so indirect formation of PFOS, PFOA, and PFNA could be an important contributor to the observed environmental burdens. A number of modeling studies have attempted to elucidate the importance of direct and indirect sources of PFOA (Armitage et al. 2006; Schenker et al. 2008; Wallington et al. 2006; Wania 2007). The study of Armitage et al. (2006) considers only direct sources, Wallington et al. (2006) and Schenker et al. (2008) consider only indirect sources, and Wania (2007) considers the impacts of both. Fluxes of PFOA through direct emissions and transport through ocean currents are estimated to be 8–22 t/year by Armitage et al. (2006) and 9–20 t/year by Wania (2007). Predicted surface ocean concentrations as a result of direct emissions (Armitage et al. 2006) were observed to be in reasonable agreement with, though slightly higher than, measurements in the North and Greenland Seas (Caliebe et al. 2005). These predicted concentrations were shown to be most sensitive to the emission values entered into the model and led to varied results depending on assumptions made regarding these emissions. All three studies in which indirect formation was examined included the impact of FTOHs; Schenker et al. (2008) also included formation from NAFSEs. These studies excluded numerous precursors discussed in Section 3, most notably FTOs, which are estimated to be present in consumer products at levels equal to FTOHs (Prevedouros et al. 2006). As a result, indirect formation is probably underestimated in all three studies. A PFOA flux to the Arctic from precursor degradation of 400 kg/year was determined by Wallington et al. (2006), while fluxes for 2005 of 154 kg/year and 113–226 kg/year were determined by Wania (2007) and Schenker et al. (2008), respectively. All three estimates are in reasonable agreement with a PFOA flux estimated from measurements of 271 kg/year for 2005 (Young et al. 2007), though the results of Wallington et al. (2006) are higher than those determined in the other two studies. Differences in results appear to reflect differing assumptions regarding emissions of FTOHs as well as the level of atmospheric chemistry included. Although Wallington et al. (2006) used FTOH emissions (1,000 t/year) that would be required to maintain the observed atmospheric FTOH concentrations, Wania (2007) and Schenker et al. (2008) included emission estimates provided by industry ($\leq$200 t/year) that are not available to the general public and are not verifiable. The atmospheric chemistry included in the models varies dramatically; Schenker et al. (2008) use a simplified chemical mechanism and Wania (2007) uses zonally averaged yields of PFOA from FTOHs. Wallington et al. (2006) include a full chemical mechanism of FTOH degradation, with measured kinetics included where available. It should be noted that the ability of perfluoroacyl radicals to lose CO (reaction 58b) was not recognized when the Wallington et al. (2006) model was developed, which may have led to an underestimation of the PFOA flux. These broad differences among the models of Wallington et al. (2006), Wania (2007), and Schenker et al. (2008) may serve to explain the discrepancy in calculated fluxes from indirect sources. Although predicted fluxes of directly emitted PFOA are higher than those of indirectly formed and deposited PFOA, these estimated numbers do not necessarily reflect the relative importance of each process with respect to biota and human exposure. More measurements are required to determine accurate fluxes and to elucidate source apportionment.

The importance of direct and indirect sources of PFNA has been examined in few studies. A single estimate of PFNA flux to the Arctic from direct sources (2–9 t/year) was determined (Armitage et al. 2009b), based on production estimates given in Prevedouros et al. (2006). The importance of PFNA deposition to the Arctic from FTOHs was studied by Wallington et al. (2006). Although a specific flux was not given, the calculated molar yield was similar to PFOA, suggesting a flux on the same order (approximately 400 kg/year). This value is in reasonable agreement with a measurement from Arctic ice of 295 kg/year for 2005 (Young et al. 2007). Again, the relative fluxes estimated for direct and indirect sources are influenced by emission estimates and require validation from further monitoring data.

Compounds require decades to travel through the ocean to reach remote areas, such as the Arctic (Li and Macdonald 2005), while transport through the atmosphere is much faster, on the order of days to weeks. Directly formed PFOS moves through the ocean, while the indirect source of PFOS, NAFSEs, moves through the atmosphere. Phase out of PFOS and all its precursors by the largest manufacturer occurred in 2001–2002. Response in environmental levels to this production change could be indicative of the relative importance of direct and indirect sources. Evidence from the western Canadian Arctic revealed that levels of PFOS have declined rapidly in recent years in both biota (Butt et al. 2007b) and snow (Young et al. 2007). Such a fast decrease in concentration suggests the importance of atmospheric sources to the Arctic. Elucidating the importance of direct and indirect sources of PFOS to the Arctic has been attempted in a single modeling study

(Armitage et al. 2009c). Direct emissions were found to account for the vast majority of PFOS found in the Arctic aqueous environment, in contrast to the hypothesis put forth by Butt et al. (2007b) to explain the temporal trend observed in ringed seals. Armitage et al. (2009c) suggest that the temporal trend could instead be caused by a change in direct uptake of precursors by biota, rather than a result of atmospheric oxidation. It should be noted that Armitage et al. (2009c) assumed that PFOS yields were equivalent to PFBS yields from smog chamber studies of NMeFBSE. As discussed above, these are unlikely to be applicable to the real environment. More work is needed to further elucidate relative sources of PFOS that occur in the remote environment.

Further evidence of the importance of atmospheric sources to remote regions is the observation of PFOS, PFOA, and PFNA appearing in areas impacted only by atmospheric sources, such as high-altitude ice caps (Young et al. 2007) and land-locked lakes (Stock et al. 2007).

Oceanic transport of directly produced PFOS, PFOA, and PFNA and atmospheric transport of precursors and subsequent degradation and deposition are both sources of these compounds to the environment. At present, it is difficult to accurately determine the relative importance of each of these processes to environmental contamination of PFOS, PFOA, and PFNA.

5.3 Long-Chained Perfluorocarboxylic Acids (PFCAs)

Sources of PFCA having chain lengths equal to or greater than 10 carbons have received relatively less attention, compared to that of PFOA, PFNA, and PFOS. None of these long-chain compounds have ever been intentionally produced in large quantities. They have been observed as impurities in a commercial product containing PFNA as the primary component (Prevedouros et al. 2006). However, it is not known how representative this single measurement is of global production. Long-chain PFCAs can also be formed from volatile precursors, as indicated in Table 19. Precipitation collected at mid-latitudes has been observed to contain PFCAs up to 12 carbons in length (Scott et al. 2006), probably as a result of atmospheric formation. Observations of PFCAs from perfluorodecanoic acid (PFDA) to perfluoropentadecanoic acid (PFPA) in Arctic wildlife (Butt et al. 2007a; Martin et al. 2004; Verreault et al. 2007) suggest that these compounds are reaching remote regions and are being bioaccumulated.

The authors of a single modeling study have examined the propensity for contamination of the Arctic by direct production of long-chain PFCAs (Armitage et al. 2009b). Using emissions determined by extrapolating the measurement of long-chain impurities of the single-product measurement reported by Prevedouros et al. (2006), fluxes of perfluoroundecanoic acid (PFUnA) and perfluorotridecanoic acid (PFTrA) were calculated as 200–1,400 and 4–145 kg/year, respectively.

Melted ice samples from the Arctic showed approximately equal levels of PFOA and PFNA (each at ~140 pg/L), as well as PFDA and PFUnA (each at ~20 pg/L)

(Young et al. 2007). Given the negligible fluxes predicted from direct production of PFDA (Armitage et al. 2009b), these data cannot be explained by direct production of long-chained PFCAs. However, these observations are consistent with indirect production from volatile precursors. Results from smog chamber studies have shown approximately equal formation of PFCA products that are one and two carbons shorter than the original fluorotelomer chain length (e.g., 8:2 FTOH produces PFNA and PFOA in approximately equal yield) (Ellis et al. 2004). Thus, similar levels of PFOA and PFNA could be explained as a result of degradation of 8:2 fluorotelomer compounds, while similar levels of PFDA and PFUnA could be a product of degradation of 10:2 fluorotelomer compounds. Further support for the importance of indirect sources is derived from the distinctive even–odd pattern observed in Arctic biota (Martin et al. 2004). Assuming approximately equal exposure to a given even chain-length PFCA and odd chain-length congener one carbon longer, a higher concentration of the odd PFCA is observed in biota due to higher bioaccumulation of the longer-chain-length PFCA. This pattern has been observed in virtually all arctic biota (Butt et al. 2007a; Houde et al. 2006; Martin et al. 2004).

Observations and trends of long-chain PFCAs are strongly indicative of volatile precursors being important sources of these compounds to the environment.

6 Summary

Perfluorocarboxylic acids (PFCAs) can be formed from the hydrolysis of perfluoroacyl fluorides and chlorides, which can be produced in three separate ways in the atmosphere. Alternatively, PFCAs can be formed directly in the gas phase through reaction of perfluoroacyl peroxy radicals or perfluorinated aldehyde hydrates. All five mechanisms have been elucidated using smog chamber techniques. Yields of PFCAs from this process vary from less than 10% to greater than 100%, depending on the mechanism. The formation of perfluorosulfonic acids in the atmosphere can also occur, though the mechanism has not been entirely elucidated. A large number of compounds have been confirmed as perfluorinated acid precursors, including CFC-replacement compounds, anesthetics, fluorotelomer compounds, and perfluorosulfonamides. Levels of some of these compounds have been measured in the atmosphere, but concentrations for the majority have yet to be detected. It is clear that atmospheric oxidation of volatile precursors contributes to the overall burden of PFAs, though the extent to which this occurs is compound and environment dependent and is difficult to assess accurately.

References

3M Company (1999) Fluorochemical use, distribution and release overview. Public Docket AR226-0550, United States Environmental Protection Agency, St. Paul, MN

3M Company (2000) Phase-out plan for POSF-based products. Public Docket OPPT-2002-0043-0009, US Environmental Protection Agency, St. Paul, MN

Acerboni G, Beukes JA, Jensen NR, Hjorth J, Myhre G, Nielsen CJ, Sundet JK (2001) Atmospheric degradation and global warming potentials of three perfluoroalkenes. Atmos Environ 35: 4113–4123

Al-Ghanem S, Battah AH, Salhab AS (2008) Monitoring of volatile anesthetics in operating room personnel using GC–MS. Jordan Med J 42(1):13–19

Antoniotti P, Borocci S, Giordani M, Grandinetti F (2008) Cl-initiated oxidation of N-ethyl-perfluoroalkanesulfonamides: a theoretical insight into experimentally observed products. J Mol Struc-Theochem 857:57–65

Armitage J, Cousins IT, Buck RC, Prevedouros K, Russell MH, Macleod M, Korzeniowski SH (2006) Modeling global-scale fate and transport of perfluorooctanoate emitted from direct sources. Environ Sci Technol 40(22):6969–6975

Armitage JM, Macleod M, Cousins IT (2009a) Modeling the global fate and transport of perfluorooctanoic acid (PFOA) and perfluorooctanoate (PFO) emitted from direct sources using a multispecies mass balance model. Environ Sci Technol 43:1134–1140

Armitage JM, Macleod M, Cousins IT (2009b) Comparative assessment of the global fate and transport pathways of long-chain perfluorocarboxylic acids (PFCAs) and perfluorocarboxylates (PFCs) emitted from direct sources. Environ Sci Technol 43:5830–5836

Armitage JM, Schenker U, Scheringer M, Martin JW, Macleod M, Cousins IT (2009c) Modeling the global fate and transport of perfluorooctane sulfonate (PFOS) and precursor compounds in relation to temporal trends in wildlife exposure. Environ Sci Technol 43:9274–9280

Atkinson R, Baulch DL, Cox RA, Crowley JN, Hampson RF, Hynes RG, Jenkin ME, Rossi MJ, Troe J, Wallington TJ (2008) Evaluated kinetic and photochemical data for atmospheric chemistry: vol IV – gas phase reactions of organic halogen species. Atmos Chem Phys 8:4141–4496

Barber JL, Berger U, Chaemfa C, Huber S, Jahnke A, Temme C, Jones KC (2007) Analysis of per- and polyfluorinated alkyl substances in air samples from Northwest Europe. J Environ Monitor 9:530–541

Barry J, Sidebottom H, Treacy J, Franklin J (1995) Kinetics and mechanism for the atmospheric oxidation of 1,1,2-trifluoroethane (HFC 143). Int J Chem Kinet 27:27–36

Bednarek G, Ereil M, Hoffmann A, Kohlmann JP, Mörs V (1996) Rate mechanism of the atmospheric degradation of 1,1,1,2-tetrafluoroethane (HFC-134a). Phys Chem Chem Phys 100(5):528–539

Bilde M, Wallington TJ, Ferronato C, Orlando JJ, Tyndall GS, Estupiñan E, Haberkorn S (1998) Atmospheric chemistry of CH$_2$BrCl, CHBrCl$_2$, CHBr$_2$Cl, CF$_3$CHBrCl, and CBr$_2$Cl$_2$. J Phys Chem A 102:1976–1986

Boulanger B, Peck AM, Schnoor JL, Hornbuckle KC (2005) Mass budget of perfluorooctane surfactants in Lake Ontario. Environ Sci Technol 39:74–79

Boutonnet JC, Bingham P, Calamari D, de Rooij C, Franklin J, Kawano T, Libre J-M, McCulloch A, Malinverno G, Odom JM, Rusch GM, Smythe K, Sobolev I, Thompson R, Tiedje JM (1999) Environmental risk assessment of trifluoroacetic acid. Hum Ecol Risk Assess 5(1): 59–124

Braun WF, Fahr A, Klein R, Kurylo MJ, Huie RE (1991) UV gas and liquid phase absorption cross section measurements of hydrochlorofluorocarbons HCFC-225ca and HCFC-225cb. J Geophys Res 96(D7):13009–13015

Brown AC, Canosa-Mas CE, Parr AD, Wayne RP (1990) Laboratory studies of some halogenated ethanes and ethers: measurements of rates of reaction with OH and of infrared absorption cross sections. Atmos Environ 24A:2499–2511

Burns DC, Ellis DA, Li X, McMurdo CJ, Webster E (2008) Experimental pK_a determination for perfluorooctanoic acid (PFOA) and the potential impact of pK_a concentration dependence on laboratory-measured partitioning phenomena and environmental modeling. Environ Sci Technol 42:9283–9288

Butt CM, Mabury SA, Muir DCG, Braune BM (2007a) Prevalence of long-chained perfluorinated carboxylates in seabirds from the Canadian Arctic. Environ Sci Technol 41:3521–3528

Butt CM, Muir DCG, Stirling I, Kwan M, Mabury SA (2007b) Rapid response of Arctic ringed seals to changes in perfluoroalkyl production. Environ Sci Technol 41(1):42–49

Butt CM, Young CJ, Mabury SA, Hurley MD, Wallington TJ (2009) Atmospheric chemistry of 4:2 fluorotelomer acrylate ($C_4F_9CH_2CH_2OC(O)CH=CH_2$): kinetics, mechanisms and products of chlorine atom and OH radical initiated oxidation. J Phys Chem A 113:3155–3161

Caliebe C, Gerwinski W, Hühnerfuss H, Theobald N (2005) Occurrence of perfluorinated organic acids in the water of the North Sea and Arctic North Atlantic in *Fluoros*, Toronto, Canada. http://www.chem.utoronto.ca/symposium/fluoros/pdfs/ANA010Theobald.pdf

Carr S, Treacy JJ, Sidebottom HW, Connell RK, Canosa-Mas CE, Wayne RP, Franklin J (1994) Kinetics and mechanisms for the reaction of hydroxyl radicals with trifluoroacetic acid under atmospheric conditions. Chem Phys Lett 227:39–44

Chan CY, Tang JH, Li YS, Chan LY (2006) Mixing ratios and sources of halocarbons in urban, semi-urban and rural sites of the Pearl River Delta, South China. Atmos Environ 40:7331–7345

Chen L, Fukuda K, Takenaka N, Bandow H, Maeda Y (2000) Kinetics of the gas-phase reaction of $CF_3CF_2CH_2OH$ with OH radicals and its atmospheric lifetime. Int J Chem Kinet 32:73–78

Chen L, Tokuhashi K, Kutsuna S, Sekiya A (2004) Rate constants for the gas-phase reaction of $CF_3CF_2CF_2CF_2CF_2CHF_2$ with OH radicals at 250–430 K. Int J Chem Kinet 36:26–33

Cheng J, Psillakis E, Hoffman MR, Colussi AJ (2009) Acid dissociation versus molecular association of perfluoroalkyl oxoacids: environmental implications. J Phys Chem A 113:8152–8156

Chiappero MS, Malanca FE, Arguello GA, Wooldridge ST, Hurley MD, Ball JC, Wallington TJ, Waterland RL, Buck RC (2006) Atmospheric chemistry of perfluoroaldehydes ($C_xF_{2x+1}CHO$) and fluorotelomer aldehydes ($C_xF_{2x+1}CH_2CHO$): quantification of the important role of photolysis. J Phys Chem A 110:11944–11953

Chiappero MS, Arguello GA, Hurley MD, Wallington TJ (2008) Atmospheric chemistry of $C_8F_{17}CH_2CHO$: yield from $C_8F_{17}CH_2CH_2OH$ (8:2 FTOH) oxidation, kinetics and mechanisms of reactions with Cl atoms and OH radicals. Chem Phys Lett 461:198–202

Clyne MAA, Holt PM (1979) Reaction kinetics involving ground $^2\Pi$ and excited $A^2\Sigma+$ hydroxyl radicals. Part 2: Rate constants for reactions of OH $^2\Pi$ with halogenomethanes and halogenoethanes. J Chem Soc Faraday T 2(75):582–591

Culbertson JA, Prins JM, Grimsrud EP, Rasmussen RA, Khalil MAK, Shearer MJ (2004) Observed trends for CF_3-containing compounds in background air at Cape Meares, Oregon, Point Barrow, Alaska, and Palmer Station, Antarctica. Chemosphere 55:1109–1119

D'eon JC, Hurley MD, Wallington TJ, Mabury SA (2006) Atmospheric chemistry of N-methyl perfluorobutane sulfonamidoethanol, $C_4F_9SO_2N(CH_3)CH_2CH_2OH$: kinetics and mechanism of reaction with OH. Environ Sci Technol 40(6):1862–1868

De Silva AO, Muir DCG, Mabury SA (2009) Distribution of perfluorocarboxylate isomers in select samples from the North American environment. Environ Toxicol Chem 28(9):1801–1814

DeMore WB (1992) Rates of hydroxyl reactions with some HFCs. P Soc Photo-Opt Ins 1715: 72–77

DeMore WB (1993a) Rate constants for the reactions of OH with HFC-134a (CF_3CH_2F) and HFC-134 (CHF_2CHF_2). Geophys Res Lett 20(13):1359–1362

DeMore WB (1993b) Rates of the hydroxyl radical reactions with some HFCs. Paper presented at the proceedings of the SPIE, The International Society for Optical Engineering

Dinglasan-Panlilio MJA, Mabury SA (2006) Significant residual fluorinated alcohols present in various fluorinated materials. Environ Sci Technol 40(5):1447–1453

Dobe S, Kachatryan LA, Berces T (1989) Kinetics of reactions of hydroxyl radicals with a series of aliphatic aldehydes. Ber Bunsen-Ges Phys Chem 93(8):847–952

Dreyer A, Temme C, Sturm R, Ebinghaus R (2008) Optimized method avoiding solvent-induced response enhancement in the analysis of volatile and semi-volatile polyfluorinated alkylated compounds using gas chromatography–mass spectrometry. J Chromatogr A 1178:199–205

Dreyer A, Ebinghaus R (2009) Polyfluorinated compounds in ambient air from ship- and land-based measurements in northern Germany. Atmos Environ 43:1527–1535

Dreyer A, Matthias V, Temme C, Ebinghaus R (2009a) Annual time series of air concentrations of polyfluorinated compounds. Environ Sci Technol 43:4029–4036

Dreyer A, Weinberg I, Temme C, Ebinghaus R (2009b) Polyfluorinated compounds in the atmo-spheric of the Atlantic and Southern Oceans: evidence for a global distribution. Environ Sci Technol 43:6507–6514

Edney EO, Gay Jr. BW, Driscoll DJ (1991) Chlorine initiated oxidation studies of hydrochloro-fluorocarbons: results for HCFC-123 (CF$_3$CHCl$_2$) and HCFC-141b (CFCl$_2$CH$_3$). J Atmos Chem 12:105–120

Edney EO, Driscoll DJ (1992) Chlorine initiated photooxidation studies of hydrochlorofluoro-carbons (HCFCs) and hydrofluorocarbons (HFCs): results for HCFC-22 (CHClF$_2$); HFC-41 (CH$_3$F); HCFC-124 (CClFHCF$_3$); HFC-125 (CF$_3$CHF$_2$); HFC-134a (CF$_3$CH$_2$F); HCFC-142b (CClF$_2$CH$_3$); and HFC-152a (CHF$_2$CH$_3$). Int J Chem Kinet 24:1067–1081

Ellis DA, Mabury SA, Martin JW, Muir DCG (2001) Thermolysis of fluoropolymers as a potential source of halogenated organic acids in the environment. Nature 412:321–324

Ellis DA, Martin JW, Mabury SA, Hurley MD, Sulbaek Andersen MP, Wallington TJ (2003) Atmospheric lifetime of fluorotelomer alcohols. Environ Sci Technol 37(17):3816–3820

Ellis DA, Martin JW, De Silva AO, Mabury SA, Hurley MD, Sulbaek Andersen MP, Wallington TJ (2004) Degradation of fluorotelomer alcohols: a likely atmospheric source of perfluorinated carboxylic acids. Environ Sci Technol 38(12):3316–3321

Ellis DA, Webster E (2009) Response to comment on "Aerosol enrichment of the surfactant PFO and mediation of the water–air transport of gaseous PFOA". Environ Sci Technol 43: 1234–1235

Forster P, Ramaswamy V, Artaxo P, Berntsen T, Betts R, Fahey DW, Haywood J, Lean J, Lowe DC, Myhre G, Nganga J, Prinn R, Raga G, Schulz M, Van Dorland R (2007) Changes in atmospheric constituents and in radiative forcing. In: S. Solomon et al. (eds) Climate change 2007: The physical science basis. Cambridge University Press, Cambridge

Gierczak T, Talukdar R, Vaghjiani GL, Lovejoy ER, Ravishankara AR (1991) Atmospheric fate of hydrofluoroethanes and hydrofluorochloroethanes: 1. Rate coefficients for reactions with OH. J Geophys Res 96(D3):5001–5011

Giesy JP, Kannan K (2001) Distribution of perfluorooctane sulfonate in wildlife. Environ Sci Technol 35:1339–1342

Gillotay D, Simon PC (1991) Temperature-dependence of ultraviolet absorption cross-sections of alternative chlorofluoroethanes: 2. The 2-chloro-1,1,1,2-tetrafluoro ethane – HCFC-124. J Atmos Chem 13:289–299

Goss K-U (2008) The pK_a values of PFOA and other highly fluorinated carboxylic acids. Environ Sci Technol 42:456–458

Hansen KJ, Clemen LA, Ellefson ME, Johnson HO (2001) Compound-specific, quantitative characterization of organic fluorochemicals in biological matrices. Environ Sci Technol 35(4):766–770

Hasson AS, Tyndall GW, Orlando JJ (2004) A product yield study of the reaction of HO$_2$ radicals with ethyl peroxy (C$_2$H$_5$O$_2$) acetyl peroxy (CH$_3$C(O)O$_2$) and acetonyl peroxy (CH$_3$C(O)CH$_2$O$_2$) radicals. J Phys Chem A 108:5979–5989

Hayman GD, Jenkin ME, Murrells TP, Johnson CE (1994) Tropospheric degradation chem-istry of HCFC-123 (CF$_3$CHCl$_2$): a proposed replacement chlorofluorocarbon. Atmos Environ 28(3):421–437

Hoshino M, Kimachi Y, Terada A (1996) Thermogravimetric behaviour of perfluoropolyether. J Appl Polym Sci 62:207–215

Houde M, Martin JW, Letcher RJ, Solomon KR, Muir DCG (2006) Biological monitoring of polyfluoroalkyl substances: a review. Environ Sci Technol 40(11):3463–3473

Houghton JT, Ding Y, Griggs DJ, Noguer M, van der Linden PJ, Dai X, Maskell K, Johnson CA (eds) (2001) Climate change 2001: the scientific basis: contribution of working group 1 to the third assessment report of the intergovernmental panel on climate change, Cambridge University Press, New York, NY, 881 pp

Howard CJ, Evenson KM (1976) Rate constants for the reactions of OH with ethane and some halogen substituted ethanes at 296 K. J Chem Phys 64(11):4303–4306

Howard PH, Meylan W (2007) EPA great lakes study for identification of PBTs to develop analytical methods: selection of additional PBTs – interim report, EPA contract No. EP-W-04-019

Hsu K-J, DeMore WB (1995) Rate constants and temperature dependences for the reactions of hydroxyl radical with several halogenated methanes, ethanes, and propanes by relative rate measurements. J Phys Chem 99(4):1235–1244

Hurley MD, Ball JC, Wallington TJ, Sulbaek Andersen MP, Ellis DA, Martin JW, Mabury SA (2004a) Atmospheric chemistry of 4:2 fluorotelomer alcohol: products and mechanism of Cl atom initiated oxidation. J Phys Chem A 108(26):5635–5642

Hurley MD, Ball JC, Wallington TJ, Sulbaek Andersen MP, Ellis DA, Martin JW, Mabury SA (2004b) Atmospheric chemistry of fluorinated alcohols: reaction with Cl atoms and OH radicals and atmospheric lifetimes. J Phys Chem A 108(11):1973–1979

Hurley MD, Sulbaek Andersen MP, Wallington TJ, Ellis DA, Martin JW, Mabury SA (2004c) Atmospheric chemistry of perfluorinated carboxylic acids: reaction with OH radicals and atmospheric lifetimes. J Phys Chem A 108:615–620

Hurley MD, Misner JA, Ball JC, Wallington TJ, Ellis DA, Martin JW, Mabury SA, Sulbaek Andersen MP (2005) Atmospheric chemistry of $CF_3CH_2CH_2OH$: kinetics, mechanisms and products of Cl atom and OH radical initiated oxidation in the presence and absence of NO_x. J Phys Chem A 109(43):9816–9826

Hurley MD, Ball JC, Wallington TJ, Sulbaek Andersen MP, Nielsen CJ, Ellis DA, Martin JW, Mabury SA (2006) Atmospheric chemistry of n-$C_xF_{2x+1}CHO$ (x = 1,2,3,4): fate of n-$C_xF_{2x+1}C(O)$ radicals. J Phys Chem A 110(45):12443–12447

Hurley MD, Ball JC, Wallington TJ (2007) Atmospheric chemistry of the Z and E isomers of $CF_3CF{=}CHF$: kinetics, mechanisms, and products of gas-phase reactions with Cl atoms, OH radicals, and O_3. J Phys Chem A 111:9789–9795

Jahnke A, Ahrens A, Ebinghaus R, Temme C (2007a) Urban versus remote air concentrations of fluorotelomer alcohols and other polyfluorinated alkyl substances in Germany. Environ Sci Technol 41:745–752

Jahnke A, Ahrens L, Ebinghaus R, Berger U, Barber JL, Temme C (2007b) An improved method for the analysis of volatile polyfluorinated alkyl substances in environmental air samples. Anal Bioanal Chem 387:965–975

Jahnke A, Berger U, Ebinghaus R, Temme C (2007c) Latitudinal gradient of airborne polyfluorinated alkyl substances in the marine atmosphere between Germany and South Africa (53°N–33°S). Environ Sci Technol 41(9):3055–3061

Jahnke A, Huber S, Temme C, Kylin H, Berger U (2007d) Development and application of a simplified sampling method for volatile polyfluorinated alkyl substances in indoor and environmental air. J Chromatogr A 1164:1–9

Jahnke A, Berger U (2009) Trace analysis of per- and polyfluorinated alkyl substances in various matrices – How do current methods perform? J Chromatogr A 1216:410–421

Jeong K-M, Hsu K-J, Jeffries JB, Kaufman F (1984) Kinetics of the reactions of OH with C_2H_6, CH_3CCl_3, $CH_2ClCHCl_2$, $CH_2ClCClF_2$, and CH_2FCF_3. J Phys Chem 88:1222–1226

Jordan A, Frank H (1999) Trifluoroacetate in the environment: evidence for sources other than HFC/HCFCs. Environ Sci Technol 33:522–527

Kanakidou M, Dentener FJ, Crutzen PJ (1995) A global three-dimensional study of the fate of HCFCs and HFC-134a in the troposphere. J Geophys Res 100(D5):18781–18801

Kannan K, Corsolini S, Falandysz J, Fillmann G, Kumar KS, Loganathan BG, Mohd MA, Olivero J, Van Wouwe N, Yang JH, Aldous KM (2004) Perfluorooctanesulfonate and related fluorochemicals in human blood from several countries. Environ Sci Technol 38:4489–4495

Kelly T, Bossoutrot V, Magneron I, Wirtz K, Treacy JJ, Mellouki A, Sidebottom H, Le Bras G (2005) A kinetic and mechanistic study of the reactions of OH radicals and Cl atoms with 3,3,3-trifluoropropanol under atmospheric conditions. J Phys Chem A 109:347–355

Kotamarthi VR, Rodriguez JM, Ko MKW, Tromp TK, Sze ND (1998) Trifluoroacetic acid from degradation of HCFCs and HFCs: a three-dimensional modeling study. J Geophys Res 103(D5):5747–5758

Langbein T, Sonntag H, Trapp D, Hoffmann A, Malms W, Röth E-P, Mörs V, Zellner R (1999) Volatile anaesthetics and the atmosphere: atmospheric lifetimes and atmospheric effects of halothane, enflurane, isoflurane, desflurane and sevoflurane. Brit J Anaesth 82(1):66–73

Leu G-H, Lee Y-P (1994) Temperature dependence of the rate coefficient of the reaction OH+CF$_3$CH$_2$F over the range 255–424 K. J Chin Chem Soc-TAIP 41:645–649

Li YF, Macdonald RW (2005) Sources and pathways of selected organochlorine pesticides to the Arctic and the effect of pathway divergence on HCH trends in biota: a review. Sci Total Environ 342:87–106

Liu R, Huie RE, Kurylo MJ (1990) Rate constants for the reactions of the OH radical with some hydrochlorofluorocarbons over the temperature range 270–400 K. J Phys Chem 94(8): 3247–3249

Louis F, Talhaoui A, Sawerysyn J-P, Rayez M-T, Rayez J-C (1997) Rate coefficients for the gas phase reactions of CF$_3$CH$_2$F (HFC-134a) with chlorine and fluorine atoms: experimental and ab initio theoretical studies. J Phys Chem A 45:8503–8507

Martin J-P, Paraskevopoulos G (1983) A kinetic study of the reactions of OH radicals with fluoroethanes: estimates of C–H bond strengths in fluoroalkanes. Can J Chemistry 61:861–865

Martin JW, Muir DCG, Moody CA, Ellis DA, Kwan W, Solomon KR, Mabury SA (2002) Collection of airborne fluorinated organics and analysis by gas chromatography/chemical ionization mass spectrometry. Anal Chem 74(3):584–590

Martin JW, Mabury SA, Solomon KR, Muir DCG (2003a) Dietary accumulations of perfluorinated acids in juvenile rainbow trout (*Oncorhynchus mykiss*). Environ Toxicol Chem 22(1):189–195

Martin JW, Mabury SA, Solomon KR, Muir DCG (2003b) Bioconcentration and tissue distribution of perfluorinated acids in rainbow trout (*Oncorhynchus mykiss*). Environ Toxicol Chem 22(1):196–204

Martin JW, Smithwick MM, Braune BM, Hoekstra PF, Muir DCG, Mabury SA (2004) Identification of long-chain perfluorinated acids in biota from the Canadian Arctic. Environ Sci Technol 38(2):373–380

Martin JW, Ellis DA, Mabury SA, Hurley MD, Wallington TJ (2006) Atmospheric chemistry of perfluoroalkanesulfonamides: kinetic and product studies of the OH and Cl atom initiated oxidation of N-ethyl perfluorobutanesulfonamide. Environ Sci Technol 40(3):864–872

Mashino M, Ninomiya Y, Kawasaki M, Wallington TJ, Hurley MD (2000) Atmospheric chemistry of CF$_3$CF=CF$_2$: kinetics and mechanism of its reactions with OH radicals, Cl atoms, and ozone. J Phys Chem A 104:7255–7260

McIlroy A, Tully FP (1993) Kinetic study of hydroxyl reactions with perfluoropropene and perfluorobenzene. J Phys Chem 97(3):610–614

McMurdo CJ, Ellis DA, Webster E, Butler J, Christensen RD, Reid LK (2008) Aerosol enrichment of the surfactant PFO and mediation of the water–air transport of gaseous PFOA. Environ Sci Technol 42:3969–3974

Meller R, Moortgat GK (1997) CF$_3$C(O)Cl: temperature-dependent (223–298 K) absorption cross-sections and quantum yields at 254 nm. J Photoch Photobio A 108:105–116

Mereau R, Rayez M-T, Rayez J-C, Caralp F, Lesclaux R (2001) Theoretical study on the atmospheric fate of carbonyl radicals: kinetics of decomposition reactions. Phys Chem Chem Phys 3:4712–4717

Miller BR, Weiss RF, Salameh PK, Tanhua T, Greally BR, Mühle J, Simmonds PG (2008) Medusa: a sample preconcentration and GC/MS detector system for in situ measurements of atmospheric trace halocarbons, hydrocarbons, and sulfur compounds. Anal Chem 80:1536–1545

Møgelberg TE, Sehested J, Bilde M, Wallington TJ, Nielsen OJ (1996) Atmospheric chemistry of CF$_3$CFHCF$_3$ (HFC-227ea): spectrokinetic investigation of the CF$_3$CFO$_2$•CF$_3$ radical, its reactions with NO and NO$_2$, and fate of the CF$_3$CFO•CF$_3$ radical. J Phys Chem 100:8882–8889

Montzka SA, Myers RC, Butler JH, Elkins JW, Lock LT, Clarke AD, Goldstein AH (1996) Observations of HFC-134a in the remote troposphere. Geophys Res Lett 23(2):169–172

Moortgat GK, Veyret B, Lesclaux R (1989) Kinetics of the reaction of HO$_2$ with CH$_3$C(O)O$_2$ in the temperature range 253–368 K. Chem Phys Lett 160(4):443–447

Nakayama T, Takahashi K, Matsumi Y, Toft A, Sulbaek Andersen MP, Nielsen OJ, Waterland RL, Buck RC, Hurley MD, Wallington TJ (2007) Atmospheric chemistry of $CF_3CH{=}CH_2$ and $C_4F_9CH{=}CH_2$: products of the gas-phase reactions with Cl atoms and OH radicals. J Phys Chem A 111:909–915

Nayak AK, Buckley TJ, Kurylo MJ, Fahr A (1996) Temperature dependence of the gas and liquid phase ultraviolet absorption cross sections of HCFC-123 (CF_3CHCl_2) and HCFC-142b (CH_3CF_2Cl). J Geophys Res 101(C4):9055–9062

Nelson Jr. DD, Zahniser MS, Kolb CE (1992) Chemical kinetics of the reactions of the hydroxyl radical with several hydrochlorofluoropropanes. J Phys Chem 96:249–253

Nelson Jr. DD, Zahniser MS, Kolb CE (1993) OH reaction kinetics and atmospheric lifetimes of CF_3CFHCF_3 and CF_3CH_2Br. Geophys Res Lett 20(2):197–200

Nielsen OJ, Javadi MS, Sulbaek Andersen MP, Hurley MD, Wallington TJ, Singh R (2007) Atmospheric chemistry of $CF_3CF{=}CH_2$: kinetics and mechanisms of gas-phase reactions with Cl atoms, OH radicals, and O_3. Chem Phys Lett 439:18–22

O'Doherty S, Cunnold DM, Manning A, Miller BR, Wang RHJ, Krummel PB, Fraser PJ, Simmonds PG, McCulloch A, Weiss RF, Salameh P, Porter LW, Prinn RG, Huang J, Sturrock G, Ryall D, Derwent RG, Montzka SA (2004) Rapid growth of hydrofluorocarbon 134a and hydrochlorofluorocarbons 141b, 142b, and 22 from Advanced Global Atmospheric Gases Experiment (AGAGE) observations at Cape Grim, Tasmania, and Mace Head, Ireland. J Geophys Res 109:DO6310

Olkhov RV, Smith IWM (2003) Time-resolved experiments on the atmospheric oxidation of C_2H_6 and some C_2 hydrofluorocarbons. Phys Chem Chem Phys 5(16):3436–3442

Oono S, Harada KH, Mahmoud MAM, Inoue K, Koizumi A (2008a) Current levels of airborne polyfluorinated telomers in Japan. Chemosphere 73:932–937

Oono S, Matsubara E, Harada KH, Takagi S, Hamada S, Asakawa A, Inoue K, Watanabe I, Koizumi A (2008b) Survey of airborne polyfluorinated telomers in Keihan area, Japan. Bull Environ Contam Tox 80:102–106

Oram DE, Reeves CE, Sturges WT, Penkett SA, Fraser PJ, Langenfelds RL (1996) Recent tropospheric growth rate and distribution of HFC-134a (CF_3CH_2F). Geophys Res Lett 23(15):1949–1952

Orkin VL, Khamaganov VG (1993) Determination of rate constants for reactions of some hydrohaloalkanes with OH radicals and their atmospheric lifetimes. J Atmos Chem 16:157–167

Orkin VL, Huie RE, Kurylo MJ (1997) Rate constants for the reactions of OH with HFC-245cb ($CH_3CF_2CF_3$) and some fluoroalkenes (CH_2CHCF_3, CH_2CFCF_3, CF_2CFCF_3, and CF_2CF_2). J Phys Chem A 101:9118–9124

Orlando JJ, Burkholder JB, McKeen SA, Ravishankara AR (1991) Atmospheric fate of several hydrofluoroethanes and hydrochloroethanes: 2. UV absorption cross sections and atmospheric lifetimes. J Geophys Res 96(D3):5013–5023

OSHA Directorate for Technical Support (2000) Anesthetic gases: guidelines for workplace exposures, Washington, DC

Papadimitriou VC, Papanastasiou DK, Stefanopoulos VG, Zaras AM, Lazarou YG, Papgiannakopoulos P (2007) Kinetic study of the reactions of Cl atoms with $CF_3CH_2CH_2OH$, $CF_3CF_2CH_2OH$, $CHF_2CF_2CH_2OH$, and $CF_3CHFCF_2CH_2OH$. J Phys Chem A 111: 11608–11617

Papadimitriou VC, Talukdar RK, Portmann RW, Ravishankara AR, Burkholder JB (2008) $CF_3CF{=}CH_2$ and (Z)–$CF_3CF{=}CHF$: temperature dependent OH rate coefficients and global warming potentials. Phys Chem Chem Phys 10:808–820

Piekarz AM, Primbs T, Fields JA, Barofsky DF, Simonich S (2007) Semivolatile fluorinated organic compounds in Asian and Western U.S. air masses. Environ Sci Technol 41:8248–8255

Prather M, Spivakovsky CM (1990) Tropospheric OH and the lifetimes of hydrochlorofluorocarbons. J Geophys Res 95(D11):18723–18729

Prevedouros K, Cousins IT, Buck RC, Korzeniowski SH (2006) Sources, fate and transport of perfluorocarboxylates. Environ Sci Technol 40(1):32–44

Prinn RG, Weiss RF, Fraser PJ, Simmonds PG, Cunnold DM, Alyea FN, O'Doherty S, Salameh P, Miller BR, Huang J, Wang RHJ, Hartley DE, Harth C, Steele LP, Sturrock G, Midgley PM, McCulloch A (2000) A history of chemically and radiatively important gases in air deduced from ALE/GAGE/AGAGE. J Geophys Res 105(D14):17751–17792

Rattigan OV, Wild O, Jones RL, Cox RA (1993) Temperature-dependent absorption cross-sections of CF_3COCl, CF_3COF, CH_3COF, CCl_3CHO and CF_3COOH. J Photoch Photobio A 73:1–9

Rattigan OV, Rowley DM, Wild O, Jones RL, Cox RA (1994) Mechanism of atmospheric oxidation of 1,1,1,2-tetrafluoroethane (HFC 134a). J Chem Soc Faraday T 90(13):1819–1829

Reimann S, Schaub D, Stemmler K, Folini D, Hill M, Hofer P, Buchmann B, Simmonds PG, Greally BR, O'Doherty S (2004) Halogenated greenhouse gases at the Swiss high alpine site of Jungfraujoch (3,580 m asl): continuous measurements and their use for regional European source allocation. J Geophys Res 109:D05307

Russell MH, Berti WR, Szostek B, Buck RC (2008) Investigation of the biodegradation potential of a fluoroacrylate polymer product in aerobic soils. Environ Sci Technol 42:800–807

Sawerysyn J-P, Talhaoui A, Meriaux B, Devolder P (1992) Absolute rate constants for elementary reactions between chlorine atoms and CHF_2Cl, CH_3CFCl_2, CH_3CF_2Cl and CH_2FCF_3 at 297 ± 2 K. Chem Phys Lett 198(1–2):197–199

Schenker U, Scheringer M, Macleod M, Martin JW, Cousins IT, Hungerbühler K (2008) Contribution of volatile precursor substances to the flux of perfluorooctanoate to the Arctic. Environ Sci Technol 42:3710–3716

Schneider WF, Wallington TJ, Huie RE (1996) Energetics and mechanism of decomposition of CF_3OH. J Phys Chem 100:6097–6103

Scollard DJ, Treacy JJ, Sidebottom HW, Balestra-Garcia C, Laverdet G, LeBras G, MacLeod H, Teton S (1993) Rate constants for the reactions of hydroxyl radicals and chlorine atoms with halogenated aldehydes. J Phys Chem 97:4683–4688

Scott BF, Macdonald RW, Kannan K, Fisk A, Witter A, Yamashita N, Durham L, Spencer C, Muir DCG (2005) Trifluoroacetate profiles in the Arctic, Atlantic and Pacific Oceans. Environ Sci Technol 39(17):6555–6560

Scott BF, Spencer C, Mabury SA, Muir DCG (2006) Poly and perfluorinated carboxylates in North American precipitation. Environ Sci Technol 40(23):7167–7174

Sehested J, Ellermann T, Nielsen OJ, Wallington TJ, Hurley MD (1993) UV absorption spectrum, and kinetics and mechanism of the self reaction of $CF_3CF_2O_2$ radicals in the gas phase at 295 K. Int J Chem Kinet 25:701–717

Sellevåg SR, Kelly T, Sidebottom H, Nielsen CJ (2004) A study of the IR and UV–Vis absorption cross-sections, photolysis and OH-initiated oxidation of CF_3CHO and CF_3CH_2CHO. Phys Chem Chem Phys 6:1243–1252

Shoeib M, Harner T, Ikonomou M, Kannan K (2004) Indoor and outdoor air concentrations and phase partitioning of perfluoroalkyl sulfonamides and polybrominated diphenyl ethers. Environ Sci Technol 38:1313–1320

Shoeib M, Harner T, Wilford BH, Jones KC, Zhu J (2005) Perfluorinated sulfonamides in indoor and outdoor air and indoor dust: occurrence, partitioning, and human exposure. Environ Sci Technol 39(17):6599–6606

Shoeib M, Harner T, Vlahos P (2006) Perfluorinated chemicals in the Arctic atmosphere. Environ Sci Technol 40:7577–7583

Shoeib M, Harner T, Lee SC, Lane D, Zhu J (2008) Sorbent-impregnated polyurethane foam disk for passive air sampling of volatile fluorinated chemicals. Anal Chem 80:675–682

Simmonds PG, O'Doherty S, Huang J, Prinn R, Derwent RG, Ryall D, Nickless G, Cunnold DM (1998) Calculated trends and the atmospheric abundance of 1,1,1,2-tetrafluoroethane, 1,1-dichloro-1-fluoroethane, and 1-chloro-1,1-difluoroethane using automated in-situ gas chromatography–mass spectrometry measurements recorded at Mace Head, Ireland, from October 1994 to March 1997. J Geophys Res 103(D13):16029–16037

Singh HB, Thakur AN, Chen YE, Kanakidou M (1996) Tetrachloroethylene as an indicator of low Cl atom concentrations in the troposphere. Geophys Res Lett 23(12):1529–1532

Smyth DV, Thompson RS, Gillings E (1994) Sodium trifluoroacetate: toxicity to the marine alga *Skeletonema costatum*. Report BL4980/B, Brixham Environmental Laboratory, Brixham

Solignac G, Mellouki A, Le Bras G, Barnes I, Benter T (2006a) Reaction of Cl atoms with $C_6F_{13}CH_2OH$, $C_6F_{13}CHO$ and C_3F_7CHO. J Phys Chem A 110(13):4450–4457

Solignac G, Mellouki A, Le Bras G, Barnes I, Benter T (2006b) Reaction of Cl atoms with $C_6F_{13}CH_2OH$, $C_6F_{13}CHO$, and C_3F_7CHO. J Phys Chem A 110:4450–4457

Solignac G, Mellouki A, Le Bras G, Yujing M, Sidebottom H (2007) The gas phase tropospheric removal of fluoroaldehydes ($C_xF_{2x+1}CHO$, $x = 3, 4, 6$). Phys Chem Chem Phys 9:4200–4210

Stemmler K, O'Doherty S, Buchmann B, Reimann S (2004) Emissions of the refrigerants HFC-134a, HCFC-22, and CFC-12 from road traffic: results from a tunnel study (Gubrist Tunnel, Switzerland). Environ Sci Technol 38:1998–2004

Stock NL, Lau FK, Ellis DA, Martin JW, Muir DCG, Mabury SA (2004) Polyfluorinated telomer alcohols and sulfonamides in the North American troposphere. Environ Sci Technol 38(4): 991–996

Stock NL (2007) Occurrence and fate of perfluoroalkyl contaminants in the abiotic environment. PhD thesis, University of Toronto, Toronto, ON, 252 pp

Stock NL, Furdui VI, Muir DCG, Mabury SA (2007) Perfluoroalkyl contaminants in the Canadian Arctic: evidence of atmospheric transport and local contamination. Environ Sci Technol 41:3529–3536

Strynar MJ, Lindstrom AB (2008) Perfluorinated compounds in house dust from Ohio and North Carolina, USA. Environ Sci Technol 42:3751–3756

Sulbaek Andersen MP, Hurley MD, Wallington T, Ball JC, Martin JW, Ellis DA, Mabury SA (2003a) Atmospheric chemistry of C_2F_5CHO: mechanism of the $C_2F_5C(O)O_2$ + HO_2 reaction. Chem Phys Lett 381:14–21

Sulbaek Andersen MP, Hurley MD, Wallington TJ, Ball JC, Martin JW, Ellis DA, Mabury SA, Nielsen CJ (2003b) Atmospheric chemistry of C_2F_5CHO: reaction with Cl atoms and OH radicals, IR spectrum of $C_2F_5C(O)O_2NO_2$. Chem Phys Lett 379:28–36

Sulbaek Andersen MP, Nielsen CJ, Hurley MD, Ball JC, Wallington TJ, Stevens JE, Martin JW, Ellis DA, Mabury SA (2004a) Atmospheric chemistry of n-$C_xF_{2x+1}CHO$ ($x = 1,3,4$): reaction with Cl atoms, OH radicals and IR spectra of $C_xF_{2x+1}C(O)O_2NO_2$. J Phys Chem A 108(24):5189–5196

Sulbaek Andersen MP, Stenby C, Nielsen CJ, Hurley MD, Ball JC, Wallington TJ, Martin JW, Ellis DA, Mabury SA (2004b) Atmospheric chemistry of n-$C_xF_{2x+1}CHO$ ($x = 1,3,4$): mechanism of the $C_xF_{2x+1}C(O)O_2$ + HO_2 reaction. J Phys Chem A 108(30):6325–6330

Sulbaek Andersen MP, Nielsen CJ, Hurley MD, Ball JC, Wallington TJ, Ellis DA, Martin JW, Mabury SA (2005a) Atmospheric chemistry of 4:2 fluorotelomer alcohol (n-$C_4F_9CH_2CH_2OH$): products and mechanism of Cl atom initiated oxidation in the presence of NO_x. J Phys Chem A 109(9):1849–1856

Sulbaek Andersen MP, Nielsen OJ, Toft A, Nakayama T, Matsumi Y, Waterland RL, Buck RC, Hurley MD, Wallington TJ (2005b) Atmospheric chemistry of $C_xF_{2x+1}CH=CH_2$ ($x = 1,2,4,6$, and 8): kinetics of gas-phase reactions with Cl atoms, OH radicals, and O_3. J Photoch Photobio A 176:124–128

Sulbaek Andersen MP, Toft A, Nielsen OJ, Hurley MD, Wallington TJ, Chishima H, Tonokura K, Mabury SA, Martin JW, Ellis DA (2006) Atmospheric chemistry of perfluorinated aldehyde hydrates (n-$C_xF_{2x+1}CH(OH)_2$, $x = 1,3,4$): hydration, dehydration, and kinetics and mechanism of Cl atom and OH radical initiated oxidation. J Phys Chem A 110:9854–9860

Talukdar R, Mellouki A, Gierczak T, Burkholder JB, McKeen SA, Ravishankara AR (1991) Atmospheric fate of CF_2H_2, CH_3CF_3, CHF_2CF_3, and CH_3CFCl_2: rate coefficients for reactions with OH and UV absorption cross sections of CH_3CFCl_2. J Phys Chem 95:5815–5821

Tang X, Madronich S, Wallington T, Calamari D (1998) Changes in tropospheric composition and air quality. J Photoch Photobio B 46:83–95

Taves DR (1968) Evidence that there are two forms of fluoride in human serum. Nature 217: 1050–1051

The European Parliament and the Council of the European Union (2006) Air pollution: emissions and fluorinated greenhouse gases from motor vehicle air-conditioning systems in Directive 2006/40/EC

Thompson RS, Stewart KM, Gillings E (1994) Trifluoroacetic acid: accumulation from aqueous solution by the roots of Sunflower (*Helianthus annuus*). Report BL5042/B, Brixham Environmental Laboratory, Brixham

Thompson RS, Stewart KM, Gillings E (1995) Sodium trifluoroacetate: toxicity to wheat (*Triticum aestivum*) in relation to bioaccumulation (by aqueous exposure of the roots). Report BL5473/B, Brixham Environmental Laboratory, Brixham

Tokuhashi K, Chen L, Kutsuna S, Uchimaru T, Sugie M, Sekiya A (2004) Environmental assessment of CFC alternatives rate constants for the reactions of OH radicals with fluorinated compounds. J Fluorine Chem 125:1801–1807

Tomy G, Budakowski W, Halldorson T, Helm PA, Stern GA, Friesen K, Pepper K, Tittlemier SA, Fisk AT (2004) Fluorinated organic compounds in an Eastern Arctic food web. Environ Sci Technol 38(24):6475–6481

Tromp TK, Ko MKW, Rodriguez JM, Sze ND (1995) Potential accumulation of a CFC-replacement degradation product in seasonal wetlands. Nature 376:327–330

Tuazon E, Atkinson R, Corchnoy S (1992) Rate constants for the gas-phase reactions of Cl atoms with a series of hydrofluorocarbons and hydrochlorofluorocarbons at 298 ± 2 K. Int J Chem Kinet 24(7):639–648

Tuazon EC, Atkinson R (1993a) Tropospheric transformation products of a series of hydrofluorocarbons and hydrochlorofluorocarbons. J Atmos Chem 17:179–199

Tuazon EC, Atkinson R (1993b) Tropospheric degradation products of CH_2FCF_3 (HFC-134a). J Atmos Chem 16:301–312

UNEP (2000) The Montreal Protocol on substances that deplete the ozone layer, United Nations Environmental Programme, Nairobi, Kenya

United States Environmental Protection Agency (2006) 2010/15 PFOA Stewardship Program. http://www.epa.gov/oppt/pfoa/pubs/pfoastewardship.htm

Velders GJM, Madronich S, Clerbaux C, Derwent R, Grutter M, Hauglustaine D, Incecik S, Ko M, Libre J-M, Nielsen OJ, Stordal F, Zhu T (2005) Chemical and radiative effects of halocarbons and their replacement compounds. In: B. Metz et al. (eds) IPCC/TEAP special report: safeguarding the ozone layer and the global climate system, Cambridge University Press, Cambridge

Verreault J, Berger U, Gabrielsen GW (2007) Trends of perfluorinated alkyl substances in herring gull eggs from two coastal colonies in northern Norway: 1983–2003. Environ Sci Technol 41:6671–6677

Vesine E, Bossoutrot V, Mellouki A, Le Bras G, Wenger J, Sidebottom H (2000) Kinetic and mechanistic study of OH- and Cl-initiated oxidation of two unsaturated HFCs: $C_4F_9CH=CH_2$ and $C_6F_{13}CH=CH_2$. J Phys Chem A 104:8512–8520

Wallington TJ, Hurley MD (1992) A kinetic study of the reaction of chlorine atoms with CF_3CHCl_2, CF_3CH_2F, $CFCl_2CH_3$, CF_2ClCH_3, CHF_2CH_3, CH_3D, CH_2D_2, CHD_3, CD_4, and CD_3Cl at 295 ± 2 K. Chem Phys Lett 189(4,5):437–442

Wallington TJ, Hurley MD, Ball JC, Kaiser EW (1992) Atmospheric chemistry of hydrofluorocarbon 134a: fate of the alkoxy radical CF_3CFHO. Environ Sci Technol 26(7): 1318–1324

Wallington TJ, Hurley MD (1993) A kinetic study of the reaction of chlorine and fluorine atoms with CF_3CHO at 295 ± 2 K. Int J Chem Kinet 25:819–824

Wallington TJ, Schneider WF, Worsnop DR, Nielsen OJ, Sehested J, Debruyn WJ, Shorter JA (1994) The environmental impact of CFC replacements – HFCs and HCFCs. Environ Sci Technol 28:320A–326A

Wallington TJ, Hurley MD, Fracheboud JM, Orlando JJ, Tyndall GS, Sehested J, Møgelberg TE, Nielsen OJ (1996) Role of excited CF_3CFHO radicals in the atmospheric chemistry of HFC-134a. J Phys Chem 100:18116–18122

Wallington TJ, Hurley MD, Fedotov V, Morrell C, Hancock G (2002) Atmospheric chemistry of $CF_3CH_2OCHF_2$ and $CF_3CHClOCHF_2$: kinetics and mechanisms of reaction with Cl atoms and OH radicals and atmospheric fate of $CF_3C(O^\bullet)HOCHF_2$ and $CF_3C(O^\bullet)ClOCHF_2$ radicals. J Phys Chem A 106:8391–8398

Wallington TJ, Hurley MD, Xia J, Wuebbles DJ, Sillman S, Ito A, Penner JE, Ellis DA, Martin JW, Mabury SA, Nielsen CJ, Sulbaek Andersen MP (2006) Formation of $C_7F_{15}COOH$ (PFOA) and other perfluorocarboxylic acids during the atmospheric oxidation of 8:2 fluorotelomer alcohol. Environ Sci Technol 40(3):924–930

Wania F (2007) Global mass balance analysis of the source of perfluorocarboxylic acids in the Arctic Ocean. Environ Sci Technol 41:4529–4535

Warren R, Gierczak T, Ravishankara AR (1991) A study of $O(^1D)$ reactions with CFC substitutes. Chem Phys Lett 183:403–409

Waterland RL, Hurley MD, Misner JA, Wallington TJ, Melo SML, Strong K, Dumoulin R, Castera L, Stock NL, Mabury SA (2005) Gas phase UV and IR absorption spectra of $CF_3CH_2CH_2OH$ and $F(CF_2CF_2)_xCH_2CH_2OH$ ($x = 2, 3, 4$). J Fluorine Chem 126:1288–1296

Waterland RL, Dobbs KD (2007) Atmospheric chemistry of linear perfluorinated aldehydes: dissociation kinetics of $C_nF_{2n+1}CO$ radicals. J Phys Chem A 111:2555–2562

Watson RT, Ravishankara AR, Machado G, Wagner S, Davis DD (1979) A kinetics study of the temperature dependence of the reactions of $OH(^2P)$ with CF_3CHCl_2, CF_3CHClF, and CF_2ClCH_2Cl. Int J Chem Kinet 11:187–197

Wild O, Rattigan OV, Jones RL, Pyle JA, Cox RA (1996) Two-dimensional modelling of some CFC replacement compounds. J Atmos Chem 25:167–199

Woodrow JE, Crosby DG, Mast T, Moilanen KW, Seiber JN (1978) Rates of transformation of trifluralin and parathion vapors in air. J Agr Food Chem 26:1312–1316

Yamada T, Fang TD, Taylor PH, Berry RJ (2000) Kinetics and thermochemistry of the OH radical reaction with CF3CCl2H and CF3CFClH. J Phys Chem A 104(21):5013–5022

Yamashita N, Kannan K, Taniyasu S, Horii Y, Petrick G, Gamo T (2005) A global survey of perfluorinated acids in oceans. Mar Pollut Bull 51:658–668

Yarwood G, Kemball-Cook S, Keinath M, Waterland RL, Korzeniowski SH, Buck RC, Russell MH, Washburn ST (2007) High-resolution atmospheric modeling of fluorotelomer alcohols and perfluorocarboxylic acids in the North American troposphere. Environ Sci Technol 41: 5756–5762

Young CJ, Donaldson DJ (2007) Overtone-induced degradation of perfluorinated alcohols in the atmosphere. J Phys Chem A 111:13466–13471

Young CJ, Furdui VI, Franklin J, Koerner RM, Muir DCG, Mabury SA (2007) Perfluorinated acids in Arctic snow: new evidence for atmospheric formation. Environ Sci Technol 41:3455–3461

Young CJ, Hurley MD, Wallington TJ, Mabury SA (2008) Atmospheric chemistry of 4:2 fluorotelomer iodide (n-$C_4F_9CH_2CH_2I$): kinetics and products of photolysis and reaction with OH radicals and Cl atoms. J Phys Chem A 112:13542–13548

Young CJ, Hurley MD, Wallington TJ, Mabury SA (2009a) Atmospheric chemistry of CF_3CF_2H and $CF_3CF_2CF_2CF_2H$: kinetics and products of gas-phase reactions with Cl atoms and OH radicals, infrared spectra, and formation of perfluorocarboxylic acids. Chem Phys Lett 473:251–256

Young CJ, Hurley MD, Wallington TJ, Mabury SA (2009b) Atmospheric chemistry of perfluorobutenes ($CF_3CF=CFCF_3$ and $CF_3CF_2CF=CF_2$): kinetics and mechanisms of reactions with OH radicals and chlorine atoms, IR spectra, global warming potentials, and oxidation to perfluorocarboxylic acids. Atmos Environ 43:3717–3724

Zellner R, Bednarek G, Hoffmann A, Kohlmann JP, Mörs V, Saathoff H (1994) Rate and mechanism of the atmospheric degradation of $2H$-heptafluoropropane (HFC-227). Ber Bunsen-Ges Phys Chem 98:141–146

Zhang Z, Liu R, Huie RE, Kurylo MJ (1991) Rate constants for the gas phase reactions of the OH radical with $CF_3CF_2CHCl_2$ (HCFC-225ca) and CF_2ClCF_2CHClF (HCFC-225cb). Geophys Res Lett 18:5–7

Zhang Z, Huie RE, Kurylo MJ (1992) Rate constants for the reactions of OH with CH_3CFCl_2 (HCFC-141b), CH_3CF_2Cl (HCFC-142b), and CH_2FCF_3 (HFC-134a). J Phys Chem 4: 1533–1535

Zhang Z, Padmaja S, Saini RD, Huie RE, Kurylo MJ (1994) Reactions of hydroxyl radicals with several hydrofluorocarbons: the temperature dependencies of the rate constants for $CHF_2CF_2CH_2F$ (HFC-245ca), $CF_3CHFCHF_2$ (HFC-236ea), CF_3CHFCF_3 (HFC-227ea), and $CF_3CH_2CH_2CF_3$ (HFC-256ffa). J Phys Chem 98:4312–4315

Isomer Profiling of Perfluorinated Substances as a Tool for Source Tracking: A Review of Early Findings and Future Applications

Jonathan P. Benskin, Amila O. De Silva, and Jonathan W. Martin

Contents

1 Introduction

The ubiquitous detection of perfluorinated acids (PFAs) and their precursors (PFA precursors) in the global environment has led to concern over their effects in humans and wildlife. This is exacerbated by evidence of developmental toxicity (Lau et al. 2007; Apelberg et al. 2007; Fei et al. 2008), along with persistence,

J.W. Martin (✉)
Division of Analytical and Environmental Toxicology, Department of Laboratory Medicine and Pathology, University of Alberta, Edmonton, AB T6G 2G3, Canada
e-mail: jon.martin@ualberta.ca

P. de Voogt (ed.), *Reviews of Environmental Contamination and Toxicology*,
Reviews of Environmental Contamination and Toxicology 208,
DOI 10.1007/978-1-4419-6880-7_2, © Springer Science+Business Media, LLC 2010

chain length-dependent bioaccumulation potential (Houde et al. 2006), and long-range transport potential (Wallington et al. 2006; Wania 2007; Armitage et al. 2006, 2009a, b). In the over half-century of global perfluorochemical manufacturing, the two most commonly used synthetic methods have produced products with very different isomeric purities. Despite the fact that both branched and linear PFA and PFA-precursor isomers exist in the environment, quantitative analysis of these chemicals is, for the most part, still conducted by eluting all isomers together and integrating them as a single peak. This practice has continued despite the fact that emerging literature suggests that more accurate and informative data can be generated by isomer-specific analysis.

The extent to which perfluoromethyl branching patterns affect the physical, chemical, and biological properties of perfluorinated substances is of increasing scientific interest. It is hypothesized that branching patterns may affect properties such as environmental transport and degradation, partitioning, bioaccumulation, pharmacokinetics, and toxicity. It may even influence total PFA quantification, thus perhaps leading to questions about the accurateness of current human and environmental exposure assessments. Of particular focus in the current chapter is the measurement and interpretation of isomer signatures in the environment to gain new knowledge on emission sources, to differentiate between historical versus current exposure sources, or to identify direct versus indirect pathways of exposure for humans and wildlife. To do this effectively requires adequate analytical methods and a fundamental knowledge of the properties that may affect the environmental fate of individual isomers.

2 Isomer Nomenclature

PFA and PFA-precursor acronyms and empirical formulae are listed in Table 1. While a comprehensive numbering system was recently proposed for all isomers of perfluoroalkyl sulfonates and carboxylates (Rayne et al. 2008b), herein we have adopted an earlier, more rudimentary, system developed by Langlois and Oehme (2006) and modified by Benskin et al. (2007) for the limited number of isomers actually present in the commercially manufactured PFA and PFA-precursor formulations (see Section 3). Using perfluorooctane sulfonate (PFOS, Table 1) as an example, linear, perfluoroisopropyl, and t-perfluorobutyl are abbreviated as n-, iso-, and tb-PFOS, respectively. For the remaining monomethyl-branched isomers, m refers to a perfluoromethyl branch and the number preceding it indicates the carbon position on which the branch resides. Likewise, dimethyl-substituted-branched isomers are labeled as m_2 and the preceding numbers refer to the locations of the CF_3 branching points. For example, 5-perfluoromethyl-PFOS is abbreviated as $5m$-PFOS, while 5,3-perfluorodimethyl-PFOS is abbreviated as $5,3m_2$-PFOS. The same nomenclature system was adopted for perfluoroalkyl carboxylates (PFCAs); however, it should be noted that $1m$-PFCAs do not exist, since the carbon in the 1-position corresponds to the carboxylate moiety.

Table 1 Acronyms and empirical formulas for perfluoroalkyl sulfonates, sulfonamides, and carboxylates

Chemical	Formula	Acronyms	Cas no.[a]
Perfluoroalkyl sulfonates			
Perfluorobutane sulfonate	$F(CF_2)_4SO_3^-$	PFBS	375-73-5
Perfluoropentane sulfonate	$F(CF_2)_5SO_3^-$	PFPeS	2706-91-4
Perfluorohexane sulfonate	$F(CF_2)_6SO_3^-$	PFHxS	355-46-4
Perfluoroheptane sulfonate	$F(CF_2)_7SO_3^-$	PFHpS	375-92-8
Perfluorooctane sulfonate	$F(CF_2)_8SO_3^-$	PFOS	1763-23-1
Perfluorodecane sulfonate	$F(CF_2)_{10}SO_3^-$	PFDS	335-77-3
Perfluoroalkyl sulfonamides			
N-Methyl perfluorooctanesulfonamidoethanol	$F(CF_2)_8SO_2N(CH_3)(CH_2CH_2OH)$	NMeFOSE	24448-09-7
N-Ethyl perfluorooctanesulfonamidoethanol	$F(CF_2)_8SO_2N(CH_2CH_3)(CH_2CH_2OH)$	NEtFOSE	1691-99-2
Perfluorooctanesulfonamide	$F(CF_2)_8SO_2NH_2$	PFOSA	754-91-6
N-Ethyl perfluorooctanesulfonamide	$F(CF_2)_8SO_2N(CH_2CH_3)H$	NEtFOSA	4151-50-2
N-Methyl perfluorooctanesulfonamide	$F(CF_2)_8SO_2N(CH_3)H$	NMeFOSA	31506-32-8
Perfluorooctanesulfonamidoethanol	$F(CF_2)_8SO_2NH(CH_2CH_2OH)$	FOSE	10116-92-4
Perfluorooctanesulfonamidoacetate	$F(CF_2)_8SO_2NH(CH_2C(O)OH)$	FOSAA	2806-24-8
N-Ethyl perfluorooctanesulfonamidoacetate	$F(CF_2)_8SO_2N(CH_2CH_3)(CH_2C(O)OH)$	NEtFOSAA	2991-50-6
N-Methyl perfluorooctanesulfonamidoacetate	$F(CF_2)_8SO_2N(CH_3)(CH_2C(O)OH)$	NMeFOSAA	2355-31-9
Perfluorocarboxylates			
Trifluoroacetate	$F(CF_2)C(O)O^-$	TFA	2966-50-9
Perfluoropropanoate	$F(CF_2)_2C(O)O^-$	PFPrA	422-64-0
Perfluorobutanoate	$F(CF_2)_3C(O)O^-$	PFBA	375-22-4
Perfluoropentanoate	$F(CF_2)_4C(O)O^-$	PFPeA	2706-90-3
Perfluorohexanoate	$F(CF_2)_5C(O)O^-$	PFHxA	307-24-4
Perfluoroheptanoate	$F(CF_2)_6C(O)O^-$	PFHpA	375-85-9
Perfluorooctanoate	$F(CF_2)_7C(O)O^-$	PFOA	335-67-1
Perfluorononanoate	$F(CF_2)_8C(O)O^-$	PFNA	375-95-1

Table 1 (continued)

Chemical	Formula	Acronyms	Cas no.[a]
Perfluorodecanoate	$F(CF_2)_9C(O)O^-$	PFDA	335-76-2
Perfluoroundecanoate	$F(CF_2)_{10}C(O)O^-$	PFUnA	2058-94-8
Perfluorododecanoate	$F(CF_2)_{11}C(O)O^-$	PFDoA	307-55-1
Perfluorotridecanoate	$F(CF_2)_{12}C(O)O^-$	PFTrA	72629-94-8
Perfluorotetradecanoate	$F(CF_2)_{13}C(O)O^-$	PFTA	376-06-7

[a]For sulfonates and carboxylates, the CAS number is for the protonated form, e.g., $F(CF_2)_7C(O)OH$ (=PFOA)

3 Historical and Current Manufacturing Sources of Perfluoroalkyl Isomers

Although the various synthetic routes and their estimated contributions to the global environmental mass balance of perfluorochemicals have been recently reviewed (Lehmler 2005; Paul et al. 2009; Prevedouros et al. 2006), here we focus on the relevance of perfluorochemical manufacturing to isomer profiles and the implications for source tracking. The major production of perfluorochemicals has historically occurred either by Simons electrochemical fluorination (ECF) or by telomerization. Telomerization is a synthetic process that results in an isomerically pure product which retains the structure of the starting material (typically linear), whereas ECF results in a mixture of branched and linear isomers and by-products. ECF was used to produce all 3M-manufactured perfluorooctane sulfonyl fluoride (PFOSF, $C_8F_{17}SO_2F$)-based products since 1949 (Paul et al. 2009) and the majority of 3M perfluorooctanoic acid since 1947 (Prevedouros et al. 2006). Although 3M phased out their perfluorooctyl-based chemistries in 2002, the company continues to manufacture perfluorobutyl-based products by this method (Parsons et al. 2008). Telomerization, which was originally developed by DuPont (Kissa 2005), saw minor use beginning in the 1970s for the production of PFOA; however, it was not until the 2002 phase out of 3M ECF PFOA that DuPont began the large-scale manufacturing of PFOA by this alternative technique. Telomerization continues to be the dominant production method today for producing PFOA and perfluorononanoic acid (PFNA); however, Prevedouros et al. (2006) indicated that minor ECF manufacturing of ammonium perfluorooctanoate (APFO) has continued since 2002 in Asia and Europe. Although it is not clear how much ECF production continues today, ECF is generally regarded as the "historical" manufacturing process whereas telomerization represents the "current" production method (De Silva and Mabury 2004, 2006).

It is only germane to note that other synthetic routes to branched PFCAs have been reported in the patent literature, as reviewed elsewhere (Lehmler 2005). For example, liquid phase direct fluorination (LPDF) can produce minor quantities of branched isomers of perfluoroalkyl substances and this may explain the small quantities of branched isomers in PFOA purchased from supplier Sigma–Aldrich/Fluka (Steinheim, Switzerland) (Table 2). However, it is unclear what contribution, if any, these minor manufacturing sources make to global PFA loadings.

The telomerization process involves free radical addition of a starting telogen (e.g., perfluoroethyl iodide, CF_3CF_2I) with an unsaturated taxogen (e.g., tetrafluoroethylene, $CF_2=CF_2$), thereby lengthening the perfluoroalkyl moiety by units of CF_2CF_2 (Kissa 2005). The major product of this reaction is typically an eight-carbon, straight-chain perfluoroalkyl iodide which is then subjected to oxidation with oleum to form PFOA (Savu 1994) or carboxylation to form PFNA, the latter of which is used by several companies in the United States, France, and Japan (Prevedouros et al. 2006) for the manufacturing of polyvinylidene fluoride. Fluoroalkyl iodides can also be reacted to form fluorotelomer

Table 2 Principally manufactured PFOS/PFOA isomer compositions, compared to specialty chemical isomer compositions determined by ^{19}F NMR. Note that in some cases, companies listed are the suppliers and not necessarily the manufacturers

Isomer	Principally manufactured isomer profiles (wt%)		Specialty chemical isomer profiles (wt%)							
	3M ECF PFOA[a]	3M ECF PFOS[b]	PFOS-TCI[c]	PFOS-matrix[d]	PFOS-Sigma-Aldrich/Fluka[e]	T-PFOS-Wellington[f]	br-PFOS-Wellington[g]	PFOS-Sigma-Aldrich/Fluka[h]	PFOS-Oakwood[i]	PFOA-Sigma-Aldrich/Fluka[j]
Lot #	Assumed equivalent in all lots (Reagen et al. 2007)		GJ01	P15D	436098/1	TPFOS0405	brPFOSK1106	Batch # 312421000	Batch # 008577, lot # XO8M	n/a
Normal	77.6	70	67.0	68.3	78.9	68.9	78.8	82.2	72.4	98.9
Internal monomethyl	12.6	17.0	18.7	17.9	9.3	17.9	–	9.8	17.7	
Isopropyl	9.0	10.3	9.6	10.9	10.0	10.8	10.0	10.0	9.4	
Alpha	0.1	1.6	3.2	1.2	1.1	1.9	1.2	1.2	3.4	
t-butyl	0.2	0.2	0.3	0.3	0.2	0.2	0.2	0.4	0.4	
Dimethyl	0.1	0.2	1.4	1.3	0.5	0.3	0.6	0.5	–	
Total branched	22.0	29.3	33.1	31.7	21.1	31.1	21.1	21.4	30.9	1.1

n/a – not available

[a] St. Paul, MN, USA (Loveless et al. 2006)
[b] St. Paul, MN, USA (Kestner 1997)
[c] Portland, OR, USA (Arsenault et al. 2008b)
[d] Columbia, SC, USA (Arsenault et al. 2008b)
[e] Milwaukee, WI, USA, mean of $n = 2$ measurements (Arsenault et al. 2008b)
[f] Guelph, ON, Canada (Wellington Laboratories 2005)
[g] Guelph, ON, Canada (Wellington Laboratories 2007)
[h] Buchs, Switzerland (Vyas et al. 2007)
[i] West Columbia, SC, USA (Vyas et al. 2007)
[j] Steinheim, Switzerland (White et al. 2009)

olefins $(F(CF_2)_nCH=CH_2)$, alcohols $(F(CF_2)_nCH_2CH_2OH)$, and fluoroacrylate monomers $(F(CF_2CF_2)_nCH_2CH_2OC(O)CHR=CH_2)$, which are subsequently incorporated into polymeric material and/or surfactants for consumer product applications. Although telomerization retains the geometry of the starting telogen and thus produces a more isomerically pure product than ECF, telomerized products usually contain chain-length impurities, which can be both even and/or odd chain lengths and varying from 4 to 15 carbons (Prevedouros et al. 2006). Despite this, there are reports in the scientific and patent literature of odd numbered and branched-chain perfluoroalkyl iodides being produced by telomerization using branched telogens (e.g., $(CF_3)_2CI$) and single-carbon taxogens (Haszeldine 1953; Grottenmuller et al. 2003; Balague et al. 1995; Millauer 1974; Katsushima et al. 1970).

In comparison to telomerization, ECF results in numerous by-products, including branched and linear isomers of various even and odd chain lengths (Fig. 1; Table 3). This method was used by 3M for perfluorination of *n*-octanoyl halide

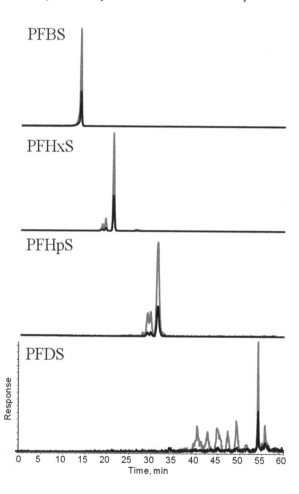

Fig. 1 LC–MS/MS chromatograms of residual perfluoroalkyl sulfonates in 3M ECF PFOS. Perfluorobutane sulfonate (PFBS; *m/z* 299/80, *grey*; *m/z* 299/99, *black*), perfluorohexane sulfonate (PFHxS; *m/z* 399/80, *grey*; *m/z* 399/99, *black*); perfluoroheptane sulfonate (PFHpS, *m/z* 449/80, *grey*; *m/z* 449/99, *black*); perfluorodecane sulfonate (PFDS, *m/z* 599/80, *grey*; *m/z* 599/99, *black*) Note increase in branched isomer content with chain length. Other sulfonates were not monitored

Table 3 Impurities and branched isomer content in 3M ECF PFOS and PFOA

Impurity in 3M ECF PFOS (lot 217)	%Impurity (wt)	%Branched of the impurity	Impurity in 3M ECF PFOA (lot 332)	%Impurity (wt)	%Branched of the impurity
PFBS	1.2[a]	0[b]	PFHxA	0.73[c]	18[d]
PFPeS	1.3[a]	N/A	PFHpA	3.7[c]	N/A
PFHxS	4.7[a]	18[e]	PFNA	0.2[f]	65[f]
PFHpS	1.1[a]	28[g]	PFDA	0.0005[f]	54[f]
PFDS	N/A	75[h]	PFUnA	0.0008[f]	28[f]
PFOA	0.79[f]	19[f]	PFDoA	0.0008[f]	32[f]
PFNA	0.002[f]	70[f]			
PFDA	0.0005[f]	51[f]			
PFUnA	0.0002[f]	46[f]			
PFDoA	0.0004[f]	33[f]			

[a] Seacat et al. (2002)
[b] As determined by LC–MS/MS peak area, monitoring m/z 299/80 transition
[c] Butenhoff et al. (2002)
[d] As determined by LC–MS/MS peak area, monitoring m/z 313/269 transition
[e] As determined by LC–MS/MS peak area, monitoring m/z 399/80 transition
[f] Reagen et al. (2007)
[g] As determined by LC–MS/MS peak area, monitoring m/z 399/80 transition
[h] As determined by LC–MS/MS peak area, monitoring m/z 599/80 transition

$(H(CH_2)_7C(O)X, X = Cl$ or $F)$ to form $F(CF_2)_7C(O)F$, which was then subjected to base-catalyzed hydrolysis to yield PFOA. The primary use of PFOA was as an emulsifier in fluoropolymer manufacturing (Prevedouros et al. 2006). Similarly, ECF of *n*-octanesulfonyl fluoride was used to produce PFOSF, which was subsequently used as a starting material for various consumer and industrial chemical formulations. For example, base-catalyzed hydrolysis of PFOSF yields PFOS, which had minor uses, predominantly in fire-fighting foams and metal plating. It was also used intentionally to some extent in various consumer products and can be observed as an unintentional residual in many PFOSF-derived products. Reaction of PFOSF with ethylamine was used to form *N*-ethyl perfluorooctanesulfonamide (NEtFOSA) (Table 1), commonly marketed as an insecticide (Appel and Abd-Elghafar 1990). The major use of PFOSF was reaction with ethyl or methyl amine, followed by ethylene carbonate, to yield *N*-ethyl perfluorooctanesulfonamidoethanol (NEtFOSE) and *N*-methyl perfluorooctanesulfonamidoethanol (NMeFOSE), respectively (Table 1). NMeFOSE was subsequently polymerized with urethane, acrylate, and/or adipate reactants to yield polymeric surface treatment products (marketed under 3M's ScotchGard[TM] brand) (3M Co. 1999). Paper protectors used in food packaging and commercial applications consisted of either NMeFOSE acrylate polymer or a mixture of 10% mono-, 85% di-, and 5% tri-phosphate esters of NEtFOSE (3M Co. 1999). It is unknown if the isomeric profile of PFOSF is preserved in subsequent consumer products that are synthetically derived from PFOSF (e.g., fluroacrylate polymers and phosphate esters). Furthermore, while the degradation of such polymers or PFOSF derivatives has been hypothesized as a source of PFAs in the environment, it is unclear whether such degradation rates would be isomer specific. The isomer profile of residual impurities may reflect the affinity of certain isomers to undergo polymerization, or alternatively, to cause selective weakening of the fluorinated polymer and cause isomer-specific degradation. Analysis of short- and long-chain perfluoroalkyl sulfonate and carboxylate impurities in standards of 3M ECF PFOA and PFOS reveal branched content of up to 75% (Table 3). Unreacted residual monomers (<1–2%) reported in polymers containing PFOSF-derived materials (3M Co. 1999; Dinglasan-Panlilio and Mabury 2006) also contain significant quantities of branched material of various chain lengths (Kissa 2005; Simons 1949).

The above discussion is important because it is uncertain what contribution residual impurities make to overall human or environmental exposures; however, it may be possible to distinguish residuals from intentionally produced products based on isomer profile. For example, residual PFOA found in pre-2002 ScotchGard[TM] Fabric and Upholstery formula as well as Rug and Carpet protectors (presumed to be NMeFOSE acrylate polymer) lacks a strong 3m-PFOA isomer signal compared to the profile of a representative 3M ECF PFOA production lot (Fig. 2). Similarly, PFOS present in ScotchGard[TM] showed enrichment of monomethyl-branched isomers when compared to 3M ECF PFOS (Fig. 2). Such differences may be useful for elucidating the role of residuals in human and environmental exposure scenarios, although further validation is necessary.

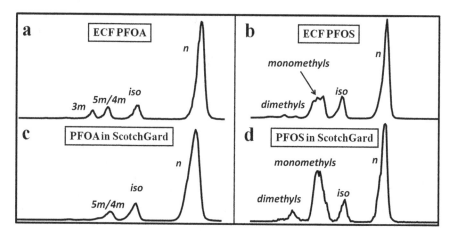

Fig. 2 Comparison of isomer profiles by LC–MS/MS of PFOA in 3M ECF standard (a), pre-2002 formula 3M ScotchGard[TM] Rug and Carpet Protector (b), PFOS in 3M ECF standard (c), and pre-2002 formula 3M ScotchGard[TM] Rug and Carpet Protector (d). For PFOA, trace represents m/z 413/369, while for PFOS, trace represents m/z 499/80. Note enrichment of iso-PFOA relative to n-PFOA in 3M ScotchGard[TM] and lack of 3m-PFOA isomer. Also notable is the apparent enrichment of PFOS monomethyl isomers relative to n-PFOS and iso-PFOS in ScotchGard[TM]. PFOS and PFOA isomer profiles in pre-2002 3M ScotchGard[TM] Rug and Carpet Protector were identical to those found in pre-2002 3M ScotchGard[TM] Fabric and Upholstery Protector

In 2002, 16 companies were known to manufacture perfluorochemicals at 33 manufacturing sites worldwide (OECD 2002; Prevedouros et al. 2006). Of these, Asahi Glass, Clariant, Daikin, and DuPont produced fluorochemicals via telomerization, while Dyneon (a subsidiary of 3M), Bayer, Dainippon Ink & Chemicals, and Miteni were known to have or are currently producing fluorochemicals by ECF (Parsons et al. 2008). Little is known about production, use, or emissions of perfluorochemicals by these manufacturers; however, it is widely reported that 3M produced 85%, or more, of total worldwide volumes of APFO by ECF since 1949 (OECD 2002). ECF manufacturing by 3M took place in plants in Cottage Grove, MN (ECF PFOA pilot production only), Cordova, IL, Decatur, AL, and Antwerp (Belgium) (3M Co. 1999). All PFOS emissions from 1951 to 1964 are assumed to have occurred in the United States, however, as production in other plants increased from 1965 to 1974, this figure decreased to 75% and by 1975 only 50% of total emissions occurred from the United States (Armitage et al. 2009a). Although the isomer composition of ECF-fluorochemicals can vary from manufacturer to manufacturer (Vyas et al. 2007) (Table 2), isomer profiles of 3M ECF PFOS and PFOA were consistent between manufacturing locations and showed minimal inter-lot variability from year to year. For example, 3M ECF PFOS reportedly had a consistent isomer composition of 70% linear (standard deviation (SD) 1.1%) and 30% branched (SD 0.8%) in eight production lots over 10 years (Reagen et al. 2007). Likewise, 3M ECF PFOA had a consistent isomer composition of 78% linear (SD 1.2%) and 22% branched (SD 1.2%) in 18 production lots over a 20-year period, as determined by ^{19}F nuclear magnetic resonance (NMR). This batch-to-batch consistency may allow researchers to distinguish sources to the environment based on isomer profiles. It is important to note from a source-tracking perspective, however, that while 3M may have produced most of the historical global ECF PFOA, between 1992 and 2002, more than 95% of 3M ECF PFOA was being used by other companies for fluoropolymer manufacturing (Wendling 2003). DuPont, for example, used 3M ECF PFOA for fluoropolymer manufacturing, beginning in the 1950s (Prevedouros et al. 2006).

Of the 89 possible PFOS isomers described by Rayne et al. (2008b), only ~11 appear to be present in measurable concentrations in 3M standards (Arsenault et al. 2008a). The structures of these isomers are n, iso, $5m$, $4m$, $3m$, $2m$, $1m$, tb, $4,4m_2$, $5,3m_2$, and $5,4m_2$. While less effort has gone into the characterization of ECF PFOA, out of 39 possible PFOA isomers (Rayne et al. 2008b) it appears that n-, iso-, $5m$-, $4m$-, and $3m$-PFOA make up 99.2% in 3M ECF standards with minor contributions (<0.8%) from $2m$, tb, $4,4m_2$, $5,3m_2$, and $5,4m_2$ (Loveless et al. 2006; Table 2). While other isomers are theoretically possible, these are unlikely to be present at measurable concentrations in the environment since they are virtually undetectable in the commercially manufactured material.

The isomer profile of 3M ECF perfluorooctane sulfonamides and sulfonamidoalcohols are also fairly consistent with 3M ECF PFOS, despite the additional synthetic production steps (Table 4). The isomer composition of these products reportedly varied from 70 to 75% straight-chain isomers, however, this could increase up to 80% linear in some cases depending on the final chemical form

Table 4 Isomer composition (relative weight %) of 3M perfluorooctane sulfonyl fluoride (PFOSF)-derived products determined by ^{19}F NMR

Isomer	FOSA[a]			PFOS[b]	NEtFOSE[a]	
	Lot 15312	TN-A-1584[c]	Lot 2353	Lot 217	Lot 30107	Mean ± 1 SD
Normal	67.3	70.9	67.1	70	69.9	69.1 ± 1.81
Monomethyl	17.9	15.2	18.2	17.0	17.4	17.2 ± 1.18
Isopropyl	9.9	9.1	9.4	10.3	10.7	9.90 ± 0.67
Alpha	3.7	3.2	3.5	1.6	1.6	2.72 ± 1.04
t-butyl	0.24	0.21	0.27	0.2	0.23	0.24 ± 0.02
Dimethyl	0.14	0.12	0.13	0.2	0.13	0.13 ± 0.01
Total branched	31.88	27.83	31.5	29.3	30.06	30.1 ± 1.63

[a] Korkowski and Kestner (1999)
[b] Kestner (1997)
[c] Lot number was not available for this standard

and customer specifications for the final product use. NMR characterization of 3M ECF perfluorooctane sulfonamide (FOSA; Table 1), PFOS, and NEtFOSE (Table 4) indicated reproducible batch-to-batch and product-to-product consistency in isomer profile. This may imply that directly emitted PFOS may be indistinguishable from precursor-derived PFOS based on isomer profile, provided the degradation pathways are not isomer selective (discussed in Section 6). This is perhaps unfortunate from a source-tracking perspective, since it may prevent the contributions of the various pathways (e.g., atmospheric transport and oxidation of precursors versus direct emission of PFAs) from being easily elucidated by isomer profile.

PFOSF-derived fluorochemicals can contribute to both perfluoro-carboxylate and sulfonate loadings via abiotic degradation (D'eon et al. 2006; Martin et al. 2006; Wallington et al. 2006; Plumlee et al. 2009) and to environmental PFOS concentrations via biotransformation (Tomy et al. 2004; Xu et al. 2004; Martin et al. 2005; Rhoads et al. 2008). Thus, PFOA isomer profiles in the environment (expected to be ∼80% linear if contribution is exclusively from ECF PFOA; Table 2) may be influenced by the isomer pattern of PFOSF-derived fluorochemicals such as perfluorooctane sulfonamides (∼70% linear; Table 4). If contributions from PFOSF-derived fluorochemicals to PFOA are significant, one might expect PFOA isomer profiles to be slightly enriched (i.e., up to 30% branched isomer content) in samples, relative to 3M ECF PFOA. However, not all branched PFOSF isomers are expected to degrade to the same corresponding branched perfluorocarboxylate. Atmospheric oxidation of alpha-branched perfluorooctyl sulfonamides (e.g. 1*m*-NEtFOSA, 1*m*-NMeFOSE) is expected to produce linear PFCAs due to loss of both the alpha carbon and its monoperfluoromethyl branch, provided degradation of branched chains proceeds via the same mechanism as the linear molecule.

Compared to what is known about the historical manufacturing of PFOS, surprisingly little is known about current production. Miteni (Italy) is known to be currently producing perfluoroalkyl sulfonates and carboxylates by ECF, and according to documents recently submitted to the International Stockholm Convention on Persistent Organic Pollutants, China began the large-scale production of PFOSF products in 2003. By 2006, 15 Chinese enterprises were producing more than 200 tonnes (t) of PFOSF, approximately half of which was exported to Brazil, the EU, and Japan (Ruisheng 2008). While this is substantially less than the 3,665 t of PFOSF produced by the 3M Co. in 2000 alone (Paul et al. 2009), it is similar to the 260 t of APFO produced by 3M in 1999 (Fluoropolymer Manufacturing Group 2002 cited in Prevedouros et al. 2006). It is not currently clear how much PFOS is being produced by China or by what method (ECF vs. telomer). If isomer profiles in new Chinese PFOSF material are unique from other manufacturers and continue to increase to pre-2002 production levels, we could expect to see changes in environmental isomer patterns in the future. To our knowledge, this "new PFOS" has yet to be taken into account in models that estimate future global PFOS production; however, Paul et al. (2009) did estimate $\sim$1,000 t/year PFOS/PFOSF manufactured globally since 2002, provided production by remaining companies has not increased.

Information recently presented at the Workshop on Managing Perfluorinated Chemicals and Transitioning to Safer Alternatives (Geneva, Switzerland, February 12–13, 2009) suggested that most manufacturers have begun substituting perfluorooctyl-based products with perfluorinated chains of four (Santoro 2009) and six (Shelton 2009; Shin-ya 2009) carbons in length. One such alternative, perfluorobutane sulfonate (PFBS), has demonstrated lower toxicity (Lieder et al. 2009) and faster elimination (Olsen et al. 2009) in rodents than its corresponding longer-chain homologs. However, PFBS is nevertheless still detectable in water, often in concentrations higher than PFOS or PFOA (Skutlarek et al. 2006; Ahrens et al. 2009). It has also been reported in children (Holzer et al. 2008), and its effects on humans are largely unknown. Residual PFBS impurities, a by-product of ECF PFOSF synthesis, did not appear to contain any branched isomers, despite the clear presence of branched C6, C7, and C10 perfluoroalkyl sulfonates in commercially manufactured material (Fig. 1). The rearrangement of the fluoroalkyl chain to form branched isomers tends to decrease with chain length (Vyas et al. 2007), and thus the lack of branched isomers in currently manufactured ECF PFBS (based on ^{19}F NMR analysis, Vyas et al. 2007), is not necessarily surprising; albeit this implies that differentiating between historical, residual, and current intentionally manufactured PFBS using isomer profiles will be difficult.

Other sources of branched PFAs may also contribute to environmental loadings. For example, thermolysis of fluoropolymers is known to be a potential source of PFAs in the environment (Ellis et al. 2001) and is thought to proceed via a carbene radical, which, while still requiring further investigation, may have the potential to form branched PFCAs. In fact, in a follow-up study by Ellis et al. (2003b), GC–MS analysis of aqueous polytetrafluoroethylene (PTFE) thermolysis extracts revealed some evidence of branched perfluorocarboxylate formation. No authentic branched standards were available at the time of this study, and therefore this finding should be re-examined using current isomer profiling methods.

4 Isomer-Specific Analytical Methodologies

4.1 Current Analytical Separation Methods

The earliest reported perfluoroalkyl isomer separations were conducted by Bastosa et al. (2001) in which GC–MS was used to partially separate a mixture of ECF NEtFOSA isomers in ant bait marketed under the name Sulfuramid™. 3M also reported HPLC–MS/MS separations of total branched from linear perfluoroalkyl carboxylates and sulfonates in human blood (Stevenson 2002) as well as ^{19}F NMR characterizations of technical mixtures (Loveless et al. 2006). These latter methods, while effective for characterizing the isomer composition in standards, lacked the sensitivity to be applied to environmental samples.

The first methods suitable for isomer-specific analysis in environmental samples were developed by De Silva and Mabury (2004), and employed GC-MS to chromatographically separate a range of perfluoroalkyl carboxylates (C8–C13). Their original method employed a 90-m ZB-35 column and was capable of separating seven PFOA isomers along with a suite of PFCAs up to and including PFTrA in less than 100 min. This method was later optimized by substituting in a 105-m Rtx-35 column (Restek Corp., Bellefonte, PA, USA), which reduced method time to under 80 min and allowed for the detection of an additional PFOA isomer (total of eight PFOA isomers) (De Silva and Mabury 2006). This has since been applied to several in vivo (De Silva et al. 2009a, c) and environmental monitoring (De Silva et al. 2009b) experiments. The advantage of GC–MS-based isomer analysis is, most obviously, not only the high chromatographic resolution associated with GC, but also that it is less prone to matrix effects which can hinder electrospray ionization (ESI) sources. Ionization efficiencies in GC–MS may have minimal differences based on comparison between pure standards of n-PFNA and iso-PFNA (De Silva and Mabury 2006). As such, the quantitative isomer composition of a sample may be possible by comparison of relative peak areas of the molecular ion. This approach is prone to errors in HPLC–ESI-MS methods because the relative peak areas are also affected by the physical properties of each isomer in the mixture. The major disadvantage of the GC–MS method is that the derivatization procedure is relatively laborious and that perfluoroalkyl sulfonates cannot be analyzed simultaneously because they are not efficiently derivatized with 2,4-difluoroaniline. Langlois and Oehme (Langlois et al. 2007) addressed this latter deficiency by developing a novel PFOS derivatization method in which PFOS was reacted with isopropanol and sulfuric acid to form volatile isopropyl derivatives. When a derivatized technical standard of PFOS was analyzed by GC–MS, 11 isomers were separated in under 8 min. This procedure proved viable for perfluorohexane sulfonate (PFHxS) and various PFCAs, however, chromatographic isomer separation was not evaluated for these latter compounds. Furthermore, it was unclear whether this method is suitable for application to environmental samples, since isomer-specific derivatization yield and stability were not investigated and no complex matrix was tested. A similar procedure producing silane derivatives for GC–MS analysis was recently used to determine PFOS and PFOA in packaging materials and textiles (Lv et al. 2009). Derivatives were noted to be stable within 2 d and limits of detection were 13.9 and 1.6 ng/mL for silated PFOS and PFOA, respectively, but the method was not validated for isomer-specific analysis. Recently, Chu and Letcher (2009) developed an in-port derivatization GC–MS method for PFOS isomers, using tetrabutyl ammonium hydroxide to produce volatile butyl PFOS isomer derivatives. Application of this method to a technical standard resulted in the separation and identification of 11 PFOS isomers in <15 min, and while detection limits were notably higher than most current LC–MS/MS methods, the authors were able to validate the method using environmental samples from the Great Lakes and Arctic.

Isomer-specific PFOS-precursor analysis by GC was also conducted by De Silva et al. (2008) and Benskin et al. (2009b) (Fig. 3). The former method employed a 60-m DB-WAX (0.25 mm ID × 0.25 μm FT, Phenomenex, Torrance, CA, USA) column to provide near-baseline resolution of nine NEtFOSE isomers in under 15 min. The latter method employed an 80-m DB35-MS column (0.25 mm ID × 0.25 μm FT, Agilent Technologies (formerly J & W Scientific), Mississauga, ON, Canada) to achieve near-baseline separation of six major and two minor NEtFOSA isomers in under 40 min.

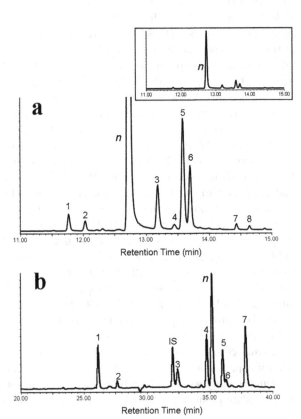

Fig. 3 a GC–(EI)MS separation of nine NEtFOSE isomers using m/z 448, 462, 562, and 540 ions on a 60-m DB-WAX. Peak #5 is tentatively identified as the isopropyl isomer.
b Solid-phase microextraction–GC-ECD separation of eight NEtFOSA isomers on an 80-m DB35-MS column. IS indicates the internal standard, n-NMeFOSA

In 2004, Martin et al. (2004) presented early chromatograms of PFOS isomer separation by HPLC–MS/MS, using an octadecasilyl (C18) stationary phase. This was followed up in 2006 by Langlois and Oehme (2006), who carried out the first isomer-specific characterization of a technical PFOS standard using purified isomer fractions and HPLC–MS/MS with perfluorophenyl (PFP) and C18 columns. This relatively fast method (<30 min with PFP column for 10 isomers), was later used to examine FOSA isomer patterns in standards, as well as PFOA isomer patterns in standards, human blood, and water (Langlois 2006). Modification of the Langlois and Oehme (2006) method has since found application in the isomer-specific analysis of PFOS in human serum/plasma (Karrman et al. 2007) and Lake Ontario biota (Houde et al. 2008) as well as long-chain perfluorinated carboxylates in Lake Ontario biota (Furdui et al. 2008). Benskin et al. (2007) later improved isomer resolution and enabled wider analyte applicability using a linear perfluorooctyl (PFO) stationary phase and an acidified mobile phase to separate and detect all the major perfluoroalkyl carboxylate, sulfonate, and sulfonamides in a single injection. While the method was comprehensive, and had the added benefit of separating out PFHxS and PFOS interferences present in human serum, it was admittedly slow (115 min). Adjustments to the gradient and equilibration conditions of this method have decreased this time to 95 min (Benskin et al. 2009a) and it has since been applied to various in vitro (Benskin et al. 2009b) and in vivo (Benskin et al. 2009a; De Silva et al. 2009a; Sharpe et al. 2010) experiments, as well as (bio)-monitoring of human blood (Riddell et al. 2009) and ocean water (Benskin et al. 2009c).

Recently, ultra-pressure liquid chromatography (UPLC) has demonstrated promise in achieving simultaneous chromatographic separations of PFOS and PFOA isomers in less than 20 min (Arsenault et al. 2008b; Riddell et al. 2009; Wellington Labs 2008). While some co-elution appears to occur between 1m- and n-PFOS, and between 4m- and iso-PFOA isomers, these can likely be resolved using knowledge of isomer-specific collision-induced dissociation patterns (Langlois and Oehme 2006; Benskin et al. 2007), as described in Section 4.3. These methods also appear to suffer from co-eluting matrix interferences, and thus some work is still needed to refine them before they can be applied to environmental samples.

Despite improvements in PFA isomer separation methods, we still lack a single method that can provide high chromatographic resolution of *all* the major PFA isomers and their interferences in a reasonable amount of time (e.g., <30 min). The recent availability of isolated and characterized standards for the major PFOS and PFOA isomers will assist greatly in the further development of quantitative isomer-specific methods. However, another existing deficiency is that commercially available technical PFOS and PFOA standards do not have the same isomer profile as those which were historically manufactured by 3M, thus making comparisons between environmental isomer profiles and historical sources of PFOA and PFOS difficult. Some researchers have obtained standards as gifts from 3M, e.g., ECF PFOS and PFOA, and while these may indeed be very useful as a "gold standard" for use in source-tracking studies, these are known to contain many impurities which make them less useful for quantitative analyses.

4.2 Analytical Quantification Bias

Martin et al. (2004) provided preliminary evidence that PFOS isomer-specific collision-induced dissociation patterns could result in an analytical bias of unknown proportion unless the isomer profile in the sample was identical to the standard used for quantification. This hypothesized bias was quantified recently by Riddell et al. (2009), in which individual purified PFOS isomers were used to compare response factors, relative to the linear isomer. These results showed that regardless of the product ion used (m/z 80 or 99), at least one PFOS isomer ($1m$-PFOS monitored using m/z 80, $4,4m_2$- and $4,5m_2$-PFOS monitored using m/z 99) will be completely absent from the chromatogram. Considering that PFOS isomer profiles in biota can vary substantially, total PFOS analysis using m/z 80 or 99 product ions will lead to some inaccuracies, and possibly, incorrect conclusions to various hypotheses. To further examine this, Riddell et al. (2009) also quantified two human serum pools containing different PFOS isomer profiles ($\sim$30–50% branched PFOS isomer content by LC–MS) using a characterized technical standard (21.1% branched PFOS by ^{19}F NMR) and isomer specific as well as total PFOS quantification methods. For sample A, total PFOS quantification resulted in m/z 80, overreporting by $\sim$30% compared to m/z 99, while quantification of sample B resulted in m/z 99, overreporting by $\sim$17% relative to m/z 80. When total branched PFOS was quantified separately from the linear isomer, the difference in values obtained from using m/z 80 and 99 for total branched isomer quantification was notably less than for total quantification methods, while consistent values were obtained for quantification of n-PFOS regardless of the product ion used (m/z 80 or 99). In the absence of methods which can quantify isomers individually, chromatographic separation of linear from "total branched" PFOS, followed by their independent quantification with a characterized technical standard, will provide improvement in the accuracy of total PFOS data.

Researchers should also be aware of a systematic bias that can be introduced when comparing isomer patterns in environmental samples at trace concentrations to ECF standards. As the concentration of branched isomers in a sample approaches the detection limit and disappears from chromatograms, the contribution of the linear isomer to total PFOA or PFOS may be incorrectly reported as 100% (Fig. 4). Any survey of isomer profiles should therefore take care to determine their "% linear dynamic range" – the concentration above which the isomer profile (or % linear calculation) of a standard stabilizes. Isomer profiles determined in samples that are below the concentration of the % linear dynamic range should only be reported with the necessary uncertainty identified, or flagged, as such. For example, Table 5 illustrates the results of Stevenson (2002), in which an ECF standard of PFOA at 0.5 and 10 ng/mL had % linear values of 81.2 and 74.7, based on LC–MS peak area, respectively, suggesting that % linear dynamic range likely lies somewhere in between these two concentrations. Alternatively, this bias may be diminished by reporting the ratio of each individually detected branched isomer to the n-isomer; thus permitting isomer-specific comparisons between studies.

Recently, the authors of several papers have utilized non-isomer-specific methods to assess the relative proportion of total branched from linear isomers in samples

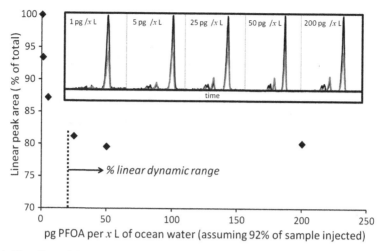

Fig. 4 The effect of decreasing concentration on PFOA isomer profile and the resulting contribution of linear PFOA to total PFOA peak area. Note that as the concentration falls outside the "% linear" dynamic range, the % contribution of linear PFOA becomes positively biased. Inset shows PFOA chromatograms for ECF standards at various concentrations. *Grey trace* represents *m/z* 413/369, *black trace* represents *m/z* 413/169

(Rylander et al. 2009a, b; Haug et al. 2009; Senthil Kumar et al. 2009). Although interesting observations have been made from this practice, caution is warranted as it can potentially lead to bias and overinterpretation of data, even when simply comparing peak areas. For example, when branched isomers are not baseline resolved from the linear isomer, it is unclear what contribution co-eluting branched isomers make to the signal of the *n*-isomer, which could potentially result in an overestimation of the true weight percent of the linear isomer in the mixture. Even in isomer-specific methods where near-baseline or baseline resolution is achieved, alpha-branched PFOS can still elute with *n*-PFOS and therefore contribute to the *m/z* 499/99 signal of this isomer (Riddell et al. 2009). For the purposes of qualitative assessment of relative branched content between samples, the above bias can be overcome by providing the corresponding branched content for a standard determined in the same manner, albeit this practice can also lead to overinterpretation (see Section 6.2), because the isomer content in standards supplied by specialty chemical manufacturers is usually not the same as historically manufactured fluorochemicals. For example, a sample with 30% branched PFOS content may appear enriched in branched isomers when compared to a Fluka standard (20% branched), despite the fact that it is indistinguishable from historically manufactured 3M ECF PFOS (30%). The numerous analytical methods (LC–MS, LC–MS/MS, and GC–MS) as well as different quantification techniques (monitoring single parent ion, single product ion, sum product ion, isomer-specific product ion) used for assessing branched isomer content can also make it difficult to compare branched content between studies. Nevertheless, this can again be overcome by providing the branched content of a characterized technical standard obtained in the same manner as the samples.

Table 5 PFOA isomer composition (%) in humans and standards. Values shown are means unless stated otherwise

References	Sample	n-	Iso-	IMM[a]	DM[b]	Other[c]	Total branched	Analysis
Stevenson (2002)	10 ng/mL 3M ECF PFOA	74.7					25.3	LC–MS/MS (C18, total ion count of m/z 369, 219, 169, 119). Branched content determined using relative sum product ions. Near-baseline separation of branched from linear
	0.5 ng/mL 3M ECF PFOA	81.2					18.8	
	Serum (Bioresource[d] Lot 020821)	99.7					<0.31	
	Serum (Lampire[e] Lot X324B)	84.0					16	
	Serum (Sigma[f] Lot 022K0965)	86.0					14	
	Serum (Golden West[g] Lot G01406042)	99.8					<0.21	
Keller et al. (2009)	Serum (SRM 1957, eight US States, 2004)	97.9					2.1	LC–MS/MS (C18 column, sum of m/z 369, 219, 169 product ions). Branched content determined using relative sum product ions. Extent of isomer separation unclear
Olsen et al. (2007)	3M ECF PFOA	78					22	LC–MS (C18, m/z 413). Branched content determined using relative peak areas. Extent of isomer separation unclear

Table 5 (continued)

References	Sample	$n-$	$Iso-$	IMM[a]	DM[b]	Other[c]	Total branched	Analysis
	Serum (occupationally exposed)	99 (range 94–99.9)					~1 (range 0.1–6.0)	
De Silva and Mabury (2006)	3M ECF PFOA	79.6	9.9	9.7		0.77	20.4	GC–MS (RTX-35, m/z 505). Branched content determined using relative peak areas. Nine PFOA isomers resolved in standard
	Serum (Sigma[f], Golden West[g])	98	1.5	0.5			2.1 (1.2–3.0)	
Benskin et al. (2007)	Serum (pregnant women, Edmonton, Canada, 2006)	>98					<2	LC–MS/MS (PFO, m/z 369). Branched content determined using relative peak areas. Eight PFOA isomers resolved in standard

[a] IMM – internal monomethyl branches ($5m$, $4m$, $3m$)
[b] DM – dimethyl branches
[c] Other – unidentified branched isomers
[d] Bioresource – Bioresource Technology Inc., Fort Lauderdale, FL
[e] Lampire – Lampire Biological Laboratories, Pipersville, PA
[f] Sigma – Sigma–Aldrich, Milwaukee, WI
[g] Golden West – Golden West Biologicals, Temecula, CA

4.3 Strategies for Isomer Separation by LC–MS/MS

For PFOS, current isomer separation techniques using PFP, PFO, or C18 station-ary phases can typically and effectively separate dimethyl-branched isomers from internal monomethyl isomers (5*m*, 4*m*, 3*m*), and *iso*-PFOS from the linear iso-mer. However, separation of individual internal monomethyl-branched isomers from each other (5*m*, 4*m*, 3*m*) on these phases often proves challenging when using only the *m/z* 99, 80, or 499 (single MS) ions. Likewise, the alpha-branch isomer (1*m*-PFOS) tends to co-elute among internal monomethyl branches on PFO (Benskin et al. 2007), with the isopropyl branch (Karrman et al. 2007) or somewhere between isopropyl and linear (Houde et al. 2008) on PFP, or with *n*-PFOS, on C18 (Riddell et al. 2009). Fortunately, these co-eluting PFOS isomers can be resolved using knowledge of isomer-specific collision-induced dissociation (Langlois and Oehme 2006) in combination with less sensitive, albeit highly specific MS/MS transitions. As shown in Fig. 5a, the *m/z* 80 product ion provides good separation of *n*, *iso*, monomethyl, and dimethyl isomers on a PFO column, similar to that which has been previously obtained on C18 (Arsenault et al. 2008b) and PFP (Houde et al. 2008;

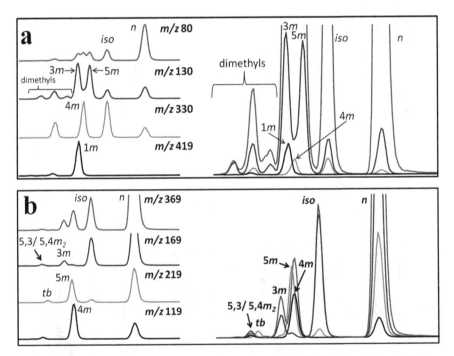

Fig. 5 **a** Recommended product ions for PFOS isomer-specific LC-MS/MS analysis: *m/z* 80 (*grey trace*): n, iso; *m/z* 130 (*dark grey trace*): 5m, 3m, dimethyls; *m/z* 330 (*light grey trace*): 4m; *m/z* 419: 1m (*black trace*). **b** Recommended product ions for PFOA isomer-specific LC-MS/MS analysis: *m/z* 369 (*grey trace*): n, iso; *m/z* 169 (*dark grey trace*): 3m, 5,3/5,4m₂; *m/z* 219 (*light grey trace*): 5m, tb; *m/z* 119 (*black trace*): 4m

Langlois and Oehme 2006) phases. With PFO, resolution of individual monomethyl branches is most easily accomplished using the m/z 130 product ion for $3m$ and $5m$ isomers and m/z 330 product ion for $4m$-PFOS. A similar strategy can be adopted for the resolution of $1m$-PFOS using m/z 419, since this is the only major isomer to produce this ion. Furthermore, $1m$-PFOS does not produce an m/z 80 ion, therefore, provided that this ion is used for quantification of iso-PFOS and n-PFOS, $1m$-PFOS should not cause any interference when using PFP or C18 phases. Although some minor isomers that are detectable in ECF PFOS standards also produce the m/z 419 ion (Fig. 5a), $1m$-PFOS is the only isomer that has been detected to date in environmental samples that produces this highly specific product ion. A similar strategy can be employed for PFOA isomers using m/z 369, 169, 219, and 119 product ions as shown in Fig. 5b. Monomethyl PFOA isomers, $5m$- and $4m$-PFOA, elute essentially together but produce distinct m/z 219 ($5m$) and 119 ($4m$) ions, respectively, which permit their resolution.

5 Influence of Physical–Chemical Properties on Environmental Fractionation of Perfluoroalkyl Isomers

Among the most intriguing topics in perfluoroalkyl research today pertains to the mechanism(s) of long-range transport of PFAs to remote regions, such as the Arctic. Much of this discussion, and the associated environmental modeling, relies heavily on accurate knowledge of physical and chemical properties. While one hypothesis suggests atmospheric transport and degradation of volatile PFA precursors (Ellis et al. 2003a; Butt et al. 2007; Young et al. 2007), another proposes slow, long-range transport of PFAs in ocean water (Armitage et al. 2006, 2009a, b; Prevedouros et al. 2006; Wania 2007). A more recent third hypothesis, presented by McMurdo et al. (2008), is that PFOA has a higher pK_a than previously thought and thus atmospheric transport of PFOA (i.e., the protonated form) may occur due to partitioning from marine aerosols, and furthermore that fractionation of branched and linear PFOA isomers may occur because of this process. The authors suggest that, based on the greater surface activity of n-PFOA (Bernett and Zisman 1967), n-PFOA will become preferentially enriched on surface microlayers. Further fractionation of linear from branched PFOA isomers would then occur during the transfer of PFOA in aerosol droplets to the gas phase due to the (presumably) differing Henry's Law constants and pK_a of all the isomers. From model calculations, pK_a values close to 0 (Goss 2008a, b), 1.3 (Lopez-Fontan et al. 2005), 2.8 (Brace 1962), and 3.8 (Burns et al. 2008) have been predicted, while values of <1 (Cheng et al. 2009) and 1.3 (Kutsuna and Hori 2008) were derived from experimental measurements. In two recent studies, pK_as of $\leq$1.5 and $\geq$3.5 (Armitage et al. 2009a; inferred from McMurdo et al. 2008), and 2.8 and 3.8 (Ellis and Webster 2009), were suggested for total branched and linear PFOAs, respectively. This was followed by computational model estimations for the pK_as of individual branched isomers of PFOA, which ranged from −0.1 ($5m$- and $4m$-PFOA) to −5.1 ($1,1,2,2m_4$-PFOA), with n-PFOA having a pK_a of −0.2

(Rayne et al. 2009). The lower values for most branched PFOA isomers are based on knowledge that electron-withdrawing CF_3 groups stabilize the carboxylate group, thus making most branched isomers stronger acids than the linear isomer. However, the inductive effect of the trifluoromethyl group is diminished as the distance from the carboxyl group increases and becomes negligible when the separation exceeds four alkyl units (Perrin et al. 1981; cited in Burns et al. 2008). Although helicity has also been suggested to influence the pK_a of n-PFOA (Burns et al. 2008), it is not clear to what extent a *lack* of helicity will influence branched isomer pK_as. Considering branching position alone, the suggestion of a significantly lower pK_a of "total branched" PFOA by Ellis et al. (2009) and Armitage et al. (2009a) requires that the majority of perfluoromethyl branches be situated alpha or beta to the carboxyl group. However, on the basis of monoperfluoromethyl isomers present in ECF PFOA, determined by ^{19}F NMR (Table 2), only 2m- and potentially 3m-PFOA have branching positions that should significantly influence the pK_a (Table 6); and the former isomer makes up only a scant 0.1% of 3M ECF PFOA. The quantities of 3m-PFOA have not been specifically reported, but total internal monomethyl branches (3m-, 4m-, and 5m-PFOA) constitute 12.6% of 3M ECF PFOA (Table 2) and 3m-PFOA is also readily identifiable in standards and in the environment. Thus, the partitioning processes described by McMurdo et al. (2008) could possibly be investigated by monitoring for a relative deficiency of 3m-PFOA in atmospheric samples over oceans or large lakes.

Overall, it is predicted that some enrichment of linear PFOA isomers may occur in the atmosphere as a result of the mechanism described by McMurdo et al. (2008). From a mass balance perspective, we speculate that the converse is unlikely and thus that ocean water PFOA isomer profiles should not be significantly influenced by isomer-specific partitioning to air. Any minor fractionation of isomers to aerosols or the atmosphere, although perhaps important as a global transport pathway, should not influence the bulk PFOA profile in the world's oceans, particularly in mid-latitude source regions. Our analyses of PFOA in Eastern Atlantic ocean water have

Table 6 Isomer composition (wt%) of ECF PFOA and predicted pK_a values

		pK_a		
	Isomer composition	Rayne et al. (2009)	Ellis and Webster (2009)	Armitage et al. (2009)
n-PFOA	77.6	−0.2	≥3.8	3.5
iso-	9	−0.1		
5m-		−0.1		
4m-	12.6[a]	−0.2	≤2.8[b]	1.5[b]
3m-		−1.3		
2m-	0.1	−1.7		
tb-	0.2	−0.3		

[a]Represents sum of 5m, 4m and 3m isomers
[b]Assumed to be average of all branched isomers

revealed a branched isomer profile (including 3m-PFOA) which was, in general, not significantly different to that of 3M ECF PFOA (Benskin et al. 2009c) (Table 7). Although this does not provide evidence *against* fractionation of PFOA isomers to marine aerosols and the atmosphere, it is strongly evident that such processes are unlikely to affect the overall isomer profiles in the world's oceans. Ultimately, more definitive evidence of selective atmospheric partitioning may be gleaned from PFOA isomer profiling in the atmosphere or in aerosols.

Other interesting differences in the physical–chemical properties of PFOA isomers imparted by perfluoroalkyl branching patterns are reported in the literature. For example, the mere ability to separate isomers by HPLC, or their derivatives by GC, implies differential hydrophobicity and vapor pressures, respectively (Langlois and Oehme 2006; Benskin et al. 2007; De Silva and Mabury 2004; Langlois et al. 2007). Vyas et al. (2007) also attributed a higher branched content in PFOSF, compared to PFOS standards, to selective removal of branched isomers during purification of the potassium salt. This is supported by the preparative-scale isolation of n-PFOS from branched isomers by successive recrystallization in water (Arsenault et al. 2008b) and carbonate buffer followed by centrifugation (Ochoa-Herrera et al. 2008), demonstrating that branched isomers are more water soluble than the linear chain – consistent with all HPLC elution orders on reversed-phase stationary phases. Furthermore, branching has also been observed to decrease melting point (Bernett and Zisman 1967), and in a separate study, to increase boiling point in perfluoroalkanes (Smart 2001).

More recently, De Silva et al. (2008) built on the work of Gauthier (2004) by measuring the n-octanol–water partition coefficients (K_{ow}) for nine isomers of 3M ECF NEtFOSE. The log K_{ow} values were statistically indistinguishable for seven branched isomers, including the isopropyl isomer, with a mean value of 5.41. However, isomer 4 (Fig. 3a) had a statistically higher K_{ow} of 5.58, and the n-isomer of NEtFOSE had a statistically lower log K_{ow} of 5.33. It is unclear whether the minor differences in these or other physical and chemical properties are sufficient to cause any significant differential transport or bioaccumulation potential. Similar studies may be warranted for other ECF chemical products and for other physical properties.

The subtle differences in the physical–chemical properties between branched and linear PFA isomers are also apparent in abiotic degradation studies. For example, Yamamoto et al. (2007) observed that three branched PFOS isomers degrade more rapidly than do the linear chain when subjected to UV light in the presence of water or alkaline 2-propanol. Similarly, Ochoa-Herrera et al. (2008) demonstrated that branched PFOS isomers could more readily undergo reductive dehalogenation by Ti(III)-citrate, in the presence of a vitamin B12 catalyst. In this study, iso- and 5m-PFOS were the most labile isomers, followed by 3m- and 4m-, 1m-, and finally n-PFOS. The authors suggested a decrease in C–C bond strength resulting from perfluorinated chain branching and/or the stabilization of radical intermediates imparted by branched structures as possible explanations for this observation. Ochoa-Herrera et al. (2008) also presented isomer-specific Gibbs free energies using ab initio calculations to predict the relative stability of the various isomers. The

Table 7 PFOA isomer composition (%) in environmental and biological samples and standards. Values shown are means unless stated otherwise

References	Sample	n-	iso-	IMM[a]	DM[b]	Other[c]	Total branched	Analysis
Benskin et al. (2009c)	3M ECF PFOA	77					23	LC–MS/MS (PFO, m/z 369, 219, 169, 119). Branched content determined using isomer-specific product ions. Ten PFOA isomers resolved in standard
	Atlantic Ocean (avg. 23 locations) (from N46° 29.386′ to N01° 13.523′ W11° 57.961′)	78					22	
	Water, coastal Asia (avg. four locations)	77					23	
De Silva and Mabury (2004)	Water, Tokyo Bay, Japan	90				2.1	10	GC–MS (ZB-35, m/z 505).
	3M ECF PFOA	77					23	Branched content determined using relative peak areas. Seven PFOA isomers resolved in standard
	Polar bear (*Ursus maritimus*) (Greenland)	95					5.0 (2.8–9.8)	
	Polar bear (*Ursus maritimus*) (Canada)	100					nd[d]	
De Silva et al. (2009b)	3M ECF PFOA	78					22	GC–MS (RTX-35 or ZB-WAX, m/z 505). Branched content determined using relative peak areas. Eight PFOA isomers resolved in standard
	Char Lake surface water (Nunavut, Canada)	99	0.69	0.39			1.1	
	Amituk Lake surface water (Nunavut, Canada)	99	0.40	0.25			0.65	
	Lake Ontario surface water (Canada)	85–94	2.8–6.8	3.1–8.5			5.9–15	
	Ontario precipitation (Canada)	96	1.9	1.7			3.7	
	Dolphin plasma (*Tursiops truncatus*) (USA)	99	0.40	0.30			0.70	

Table 7 (continued)

References	Sample	n-	iso-	IMM[a]	DM[b]	Other[c]	Total branched	Analysis
	Lake Trout (*Salvelinus namaycush*) (Lake Ontario, Canada)	95	2.9	2.1			4.9	
	Mysis (*Mysis relicta*) (Lake Ontario, Canada)	99	0.69	0.40			1.1	
	Diporeia (*Diporeia hoyi*) (Lake Ontario, Canada)	99	0.79	0.42			1.2	
	Alewife (*Alosa pseudoharengus*) (Lake Ontario, Canada)	99	0.59	0.50			1.1	
	Sculpin (*Cottus cognatus*) (Lake Ontario, Canada)	99	0.69	0.40			1.1	
	Smelt (*Osmerus mordax*) (Lake Ontario, Canada)	99	0.6	0.24			0.83	
	Zooplankton (Lake Ontario, Canada)	98–99	0.99	0.39– 0.79			1.4–1.8	
	Ringed seal (*Phoca hispida*) (Lake Ontario, Canada)	100					nd	
	Lake Ontario sediment (Canada)	97–98	0.98–1.4	0.59–1.5			1.6–2.8	
	Char Lake sediment (Nunavut, Canada)	95	2.9	2.1			4.9	
Furdui et al. (2008)	Trout (*Salvelinus namaycush*) (Lake Ontario, Canada)	100					nd	LC–MS/MS (C18 column, *m/z* 369). Branched content determined using relative peak areas. Extent of isomer separation unclear

Table 7 (continued)

References	Sample	n-	iso-	IMM[a]	DM[b]	Other[c]	Total branched	Analysis
Powley et al. (2008)	Arctic Cod (*Arctogadus glacialis*) (Canada)	100					nd	LC–MS/MS (C8 column, *m/z* 369). Branched content determined using relative peak areas. Extent of isomer separation unclear
	Bearded and ringed Seal (*Erignathus barbatus and Phoca hispida*) (Canada)	100					nd	

[a]IMM – internal monomethyl branches (5*m*, 4*m*, 3*m*)
[b]DM – dimethyl branches
[c]Other – unidentified branched isomers
nd – not detected

results indicated that after n-PFOS, $1m$- and iso-PFOS were the most stable, followed by $3m$-, $4m$-, and $5m$-PFOS. These data contrast those of Rayne et al. (2008a), in which branched isomers were all found to be more thermodynamically stable than n-PFOS; based on gas-phase enthalpies, entropies, and free energies of formation. If municipal or industrial water treatment facilities begin to apply such catalytic reductive or oxidative treatment procedures then some unique isomer profiles may be relevant in local environments.

The stability of perfluorinated radical intermediates may also influence the isomer-specific abiotic oxidation of ECF perfluoroalkyl sulfonamides. These are hypothesized to be a source of branched PFOS and perfluoroalkyl carboxylates (including PFOA) in the environment based on their occurrence in smog chamber studies with volatile precursors in the presence of Cl and OH radicals (D'eon et al. 2006; Martin et al. 2006) and indirect photolysis experiments with OH radicals (Plumlee et al. 2009). It should be noted that alpha-branched PFOA is unlikely to form from oxidation of alpha-branched perfluorooctyl sulfonamides by these processes (Ellis et al. 2004; Martin et al. 2006). Thus, while an absence of $2m$-PFOA may be indicative of oxidation processes, it is only present at trace levels in 3M ECF PFOA (<0.1%) and has yet to be detected by current LC–MS methods, perhaps due to its unique collision-induced dissociation to m/z 85 (CF_3O^-) and 63 ($[CO_2F]^-$) product ions (Benskin et al. 2009a). Isomer-specific monitoring of PFA atmospheric deposition in remote regions may provide insight into this issue.

6 Characterization of Perfluoroalkyl Isomer Profiles in the Environment

6.1 PFOA Isomer Profiles

Especially relevant for regulation of fluorochemicals is the extent of the environmental PFA burden attributed to current-use fluorochemicals versus those whose source has largely been regulated or phased out. Early on, it was hypothesized that this could be assessed by monitoring isomer profiles in biological samples, since historical (pre-2002 phase out) releases of ECF fluorochemicals consisted of a mixture of isomers whereas current and historical manufacture of telomer-derived products has largely been of strictly the linear isomer (De Silva and Mabury 2006). However, due to the preferential excretion of branched isomers, it is possible that, at steady state, tissues of organisms exposed exclusively to ECF PFOA could take on isomer profiles that are predominantly linear (Benskin et al. 2009a; De Silva et al. 2009a, c). This evidence from rodents and fish thus raises ambiguity when attempting to ascribe manufacturing source based on PFOA isomer patterns in biological samples.

Notwithstanding, some information may be gained by examining isomer profiles in biological samples. For example, De Silva and Mabury (2004) examined PFCA isomer profiles in Arctic polar bear livers, from the south eastern Hudson Bay region of Canada, and central eastern Greenland and found that Greenland bears showed some contribution from an electrochemical source (i.e., minor detectable branched isomers; Table 7), whereas Canadian bears had none detectable. Canadian bears had higher total concentrations of PFOA (mean 25 ng/g) compared to Greenland bears (9 ng/g), and thus the absence of branched isomers in Canadian bears cannot be explained by detection limits. Consistent with this observation, there were also no detectable branched PFOA isomers in seals (De Silva et al. 2009b) or cod (Powley et al. 2008) from the western and central Canadian Arctic. The discrepancy among polar bear populations may result from exposure to PFOA from two different sources or via different transport mechanisms. For example, Greenland bears may have PFOA isomer signatures similar to what is transported from the Arctic, which appear predominantly electrochemical in origin (Benskin et al. 2009c). Conversely, the strictly linear signature of Canadian polar bears may indicate less exposure to PFAs which have undergone long-range transport in oceans and more exposure to telomer-derived PFAs which have undergone atmospheric transport, since the atmosphere has been shown to deliver a highly linear profile of PFOA as evidenced by 99% n-PFOA in water and 95% n-PFOA in sediment from isolated remote Arctic lakes (De Silva et al. 2009b).

PFOA isomer profiles in samples (biotic and abiotic) from throughout North America also reveal a predominantly linear signature, albeit a "% linear" dynamic range was not defined in many of these studies, thus the % linear may be positively biased for some samples (see Section 4.2). Furdui et al. (2008) observed only n-PFOA in isomer profiles in Lake Ontario Lake Trout and suspended sediment. Consistent with this result, in a separate study De Silva et al. (2009b) observed predominantly n-PFOA (95%) in biological samples from Lake Ontario but substantially more branched PFOA isomers in surface water (85–94% linear), supporting the hypothesis of isomer-specific biological discrimination (De Silva et al. 2009c). In samples that contained branched isomers, n-, iso-, and $5m$-PFOA were detected in humans, rainwater, Lake Ontario surface water and biota, and dolphins (Table 7). Of these, Lake Ontario surface water (87–93% linear PFOA) also contained $4m$-PFOA and appeared to have the profile most similar to that of 3M ECF PFOA.

To date, the highest relative quantity of branched PFOA measured in environmental samples is in ocean water from the Atlantic and coastal Asia (Benskin et al. 2009c) (Table 7). In these samples, PFOA isomer profiles were, for the most part, consistent with a 3M ECF PFOA standard. The exception was in samples from Tokyo Bay which appeared to contain significant additional contributions from a linear (presumably telomer) source, but these samples also did not have a consistent ratio of iso-PFOA to other branched PFOA isomers, suggesting a potential additional source of iso-PFOA (also presumably telomer). This latter hypothesis is supported by the observation of single branched isomers (assumed to be isopropyl) in addition to linear isomers of PFNA, perfluorodecanoate (PFDA), perfluoroundecanoate (PFUnA), and perfluorododecanoate (PFDoA) in Toyko Bay (absent in all other coastal Asian sampling locations), as well as recent data presented by Zushi et al. (2010) in Tokyo Bay sediment cores (see Section 6.3).

In humans, the PFOA isomer signature appears predominantly linear regardless of location and sex (Table 5). Serum from the background population and from occupationally exposed men and pregnant or non-pregnant women in four different studies showed consistently ≤2% total branched PFOA content (De Silva and Mabury 2006; Olsen et al. 2007; Benskin et al. 2007; Keller et al. 2009). Interestingly, the highest relative amount of branched PFOA in human serum is from unpublished data by 3M, in which a number of pooled human serum samples were analyzed by LC–MS/MS and branched content of up to 16% was observed, compared to 25.3% branched in a 10-ng/mL 3M ECF standard (Table 7). It should be noted that the apparent elevated quantity of branched isomers in this 3M ECF standard is likely a result of simply summing the responses of m/z 119, 169, 219, and 369 product ions and is therefore not representative of the actual weight % of branched isomers in 3M ECF PFOA, which is acknowledged as ~20% (Reagen et al. 2007).

6.2 Perfluoroalkyl Sulfonate and Sulfonamide Isomer Profiles

The exclusive production of PFOS and PFOS precursors by ECF makes PFOS isomer signatures a potentially powerful tool for conducting exposure source determination experiments. Unlike PFOA, which has a predominantly linear isomer signature in humans, PFOS isomer profiles vary depending on geographic location and time of sample collection (Table 8). For example, in one of the earliest studies of PFOS isomer profiling in humans, contributions of the linear isomer to total PFOS ranged from ~59 (Australia serum and UK plasma) to ~68% (Sweden plasma) compared to a Fluka standard (78%) (Karrman et al. 2007). At the time, this apparent preferential accumulation of branched PFOS isomers was attributed to pharmacokinetic discrimination, however, this is contrary to what is observed for PFOS isomers in rodents (Benskin et al. 2009a; De Silva et al. 2009a), where the linear isomer was preferentially retained, albeit non-significantly relative to most branched isomers. Furthermore, PFOS standards manufactured by Sigma–Aldrich/Fluka (~80% n-PFOS by ^{19}F NMR; Table 2) are known today to have lower branched isomer content than 3M ECF PFOS (~70% n-PFOS by ^{19}F NMR, Table 2), thus a Fluka standard is a non-ideal reference standard, as we indicated earlier, and it is not clear if both these human samples would have been significantly different from 3M ECF PFOS. Nonetheless, the difference between the % linear values found in Australia/UK and Sweden implies that some factor, whether it be pharmacokinetic or source, is influencing the isomer profiles in these locations. Furthermore, Haug et al. (2009) also reported the apparent enrichment of branched PFOS isomers in a more recent survey of human blood samples from Norway. In this study, an 11% decrease in the relative proportion of n-PFOS was observed between 1976 (68% linear) and 2007 (57% linear), albeit the branched content in a reference standard was not provided. Interestingly, the same trend of decreasing branched content with time was also observed by Riddell et al. (2009) in human

Table 8 PFOS isomer composition (%) in humans. Values shown are means unless stated otherwise

References	Sample (description, location, year collected)	n-	iso-	5m/4m/3m	Dimethyls	Other branched	Total branched	Analysis/comment
Karrman et al. (2007)	PFOS standard (Fluka)	78.0	14.4[a]	8.0	0.6		23.0	LC–MS (C18, m/z 499). Branched content determined using relative peak areas. Separation of several branched isomers and near separation of branched from linear isomers
	Plasma (mixed age/gender, Sweden, 1997–2000)	68.1	18.0[a]	12.6	0.4	0.9	31.9	
	Serum (mixed age/gender, Australia, 2002–2003)	58.7	21.3[a]	17.1	0.8	2.9	42.1	
	Plasma (mixed age/gender, UK, 2003)	59.6	20.4[a]	17.7	0.5	2.5	41.1	
Haug et al. (2009)	Serum (*range*, mixed age/gender, Norway, 1976–2007)	53–78					22–47	LC–MS/MS (C8 column, m/z 499/499). Branched content determined using relative peak areas. Extent of isomer separation unclear
	Serum (mixed age/gender, Norway, 1976)	68					32	
	Serum (mixed age/gender, Norway, 2007)	57					43	
Keller et al. (2009)	Serum (SRM 1957, eight US States, 2004)	59					41	LC–MS/MS (C18 column, sum of m/z 80, 99, 130 product ions). Branched content determined using relative sum product ions. Extent of isomer separation unclear
	PFOS standard	77					23	LC–MS/MS (C8 column, m/z 80). Branched content estimated from peak height in chromatograms found in Keller et al. (2009). Branched isomers not baseline resolved
	Serum (SRM 1957, eight US States, 2004)	65					35	
	Milk (SRM 1954, several US states, 2006)	73					27	

Table 8 (continued)

References	Sample (description, location, year collected)	n-	iso-	5m/4m/3m	Dimethyls	Other branched	Total branched	Analysis/comment
Riddell et al. (2009)	Serum (SRM 1589a, Great Lakes region, 1996)	~70					~30	LC–MS/MS (PFO column, m/z 80). Total branched quantified separately from linear using characterized standard. Branched isomers baseline resolved from n-PFOS
	Serum (SRM 1957, eight US States, 2004)	~50					~50	
Benskin et al. (2007)	PFOS standard (Fluka)	76.0					24.0	LC–MS/MS (PFO column, m/z 80). Concentration using n-PFOS and technical PFOS (Fluka) standards. Branched isomers baseline resolved from n-PFOS
	Serum (pregnant women, Edmonton, Canada, 2006)	80.0					20.0	
Rylander et al. (2009b)	Plasma (*median*, delivering women, south central Vietnam, 2005)	83 (range 17–93)					17 (range 7–83)	LC-Q-TOF (C18, m/z 498.93). Branched content determined using relative peak areas. Branched appears to be separated from linear. Extent of co-elution with linear unclear
Rylander et al. (2009a)	Plasma (male, Norway, 2005)	67 (range 49–100)					33 (range 0–51)	LC-Q-TOF (C18, m/z 498.93). Branched content determined using relative peak areas. Branched appears to be separated from linear. Extent of co-elution with linear unclear
	Plasma (female, Norway, 2005)	70% (linear, range 56–100)					30 (range 0–44)	

[a] iso+ 1m isomer

serum standard reference materials (SRMs) collected in 1996 (SRM 1589a) and 2004 (SRM 1957) using isomer-specific quantification. SRM 1589a was collected across eight States and SRM 1957 was collected from the Great Lakes region, and it is unclear what geographical location-related factors may have influenced these profiles. Nonetheless, SRM 1589a (~30% branched) clearly had lower branched content than did SRM 1957 (~50% branched) and this was supported by the results of Keller et al. (2009), who also found SRM 1957 to contain elevated branched content (41%), although the branched content for a reference standard was not provided. Interestingly, a qualitative comparison of branched PFOS content in human serum (SRM 1957), human milk (SRM 1954), and a technical standard (unknown supplier) based on peak heights from chromatograms provided in Keller et al. (2009) indicates that the branched content in human milk may be quite similar to the reference standard and deficient in branched content relative to human serum (SRM 1957). It is unclear what factors might be influencing these isomer profiles and further investigation is needed; still, it is reasonable to hypothesize that the linear isomer, being more lipophilic, might partition to human milk to a greater extent than do branched isomers.

The results of Haug et al. (2009) are generally supported by those of Rylander et al. (2009a), who also examined human blood samples from Norway collected in 2005 and found a similar contribution of linear isomers; notwithstanding there was a small, albeit significant difference between men (67% linear, range 49–100%) and women (69% linear, range 56–100%), but again, these were not compared to a technical standard, thus it is difficult to say if these are similar to the ~70% linear content in 3M ECF PFOS or not. Isomer profiles of PFHxS, PFOS, and FOSA have also been examined in the serum of pregnant women from Edmonton, Canada (Benskin et al. 2007). Although up to six branched isomers were detected in a PFHxS standard from Fluka, endogenous interferences present in the serum (as discussed in Chan et al. 2009) hampered the elucidation of PFHxS isomer profiles, thus only a single branched PFHxS isomer was detected in addition to n-PFHxS. Two branched FOSA isomers were also observed, in addition to n-FOSA. PFOS isomer profiles in human serum were very similar (~80% linear based on quantification using an n-PFOS standard) to a Fluka standard (76%), suggesting that the branched isomer content was substantially lower than that of 3M ECF PFOS. This is generally consistent with recent data by Rylander et al. (2009b), in which the median contribution of the n-isomer to total PFOS in delivering women from south central Vietnam was 83% (range 17–93%), based on LC–MS/MS analysis; however, it is unclear if all branched isomers were fully resolved from the linear chain or how this percentage compared to that in a technical standard. Nonetheless, when taken at face value, it is interesting that samples from both of these studies were collected from pregnant or delivering women and both reported a deficiency in branched PFOS content relative to studies from Norway (Haug et al. 2009; Rylander et al. 2009a), Australia, and the UK (Karrman et al. 2007), which generally showed enrichment of branched content and did not involve pregnant or delivering women. While it is unknown what factor or combination of factors (e.g., source and pharmacokinetics) contribute to these differences, one possibility is that pregnancy reduces the body burden of branched isomers in the mother by preferentially transferring branched PFOS to the fetus. At this time this hypothesis remains tentative, however, recent data have shown that branched PFOS isomers can preferentially cross the placental barrier relative to n-PFOS (Beesoon et al. 2009).

In contrast to humans, enrichment of branched PFOS in wildlife has not been frequently observed (Table 9). Lloyd et al. (2009) qualitatively observed enrichment of branched isomers in Red deer liver and Whitebait relative to a Fluka PFOS standard, while Powley et al. (2009) reported 50% branched PFOS in cod from the western Canadian Arctic, relative to Fluka PFOS (74%). In all other literature to date, PFOS isomer profiles in wildlife appear either similar or deficient in branched content relative to technical standards. For example, Chu and Letcher (2009) observed enrichment of n-PFOS in eggs from herring gull (94.5% n-PFOS) and double-crested cormorant (95.9% n-PFOS) from the Great Lakes, as well as in polar bear samples from the Norwegian Arctic (plasma, 82.4% n-PFOS) and Canadian Arctic (liver, 92.4% n-PFOS), compared to a technical standard from Wellington (65% n-PFOS). In these samples, dimethyl-branched isomers ($3,5m_2$, $4,5m_2$, tb-, and $4,4m_2$-PFOS) were not detectable, which is consistent with the results of Houde et al. (2008), who also found an absence of dimethyl-branched isomers and enrichment of n-PFOS in a Lake Ontario foodweb. In this study, the n-isomer accounted for more than 88% total PFOS in all biological samples, which was similar to that observed in sediment (81–89% n-PFOS) but contrasted with the composition in Fluka PFOS (77%) and Lake Ontario water (43–56% n-PFOS), the latter of which was noticeably deficient in n-PFOS. Powley et al. (2008) reported similar results in bearded and ringed seals from the western Canadian Arctic, which were both highly enriched in n-PFOS (96%) compared to a Sigma–Aldrich/Fluka standard (76%). Likewise, Senthil Kumar et al. (2009), observed 77–89% n-PFOS in a range of aquatic wildlife from GA, United States, and Lloyd et al. (2009) reported a qualitative deficiency in branched PFOS in Cromer crab and Carp roe, compared to a Fluka standard.

Although the deficiency in branched PFOS observed in most wildlife may be explained by preferential absorption or retention of the linear isomer, as observed to a minor extent in rodents (Benskin et al. 2009a; De Silva et al. 2009a) and significantly in fish (Sharpe et al. 2010) (see Section 7), it is also possible that in biological samples where total PFOS concentrations are extremely low, some branched isomers may be below detection limits, resulting in a positive bias in the % of total PFOS attributed to the linear isomer (see % linear dynamic range in Section 4.2). Nevertheless, deficiencies in branched content are still observed in samples containing total PFOS concentrations which are well above isomer detection limits in ECF standards, therefore there is a reasonable degree of confidence in these data.

Interestingly, the positive analytical bias discussed above, and what is known about the pharmacokinetics of PFOS isomers, does not explain the frequent observation of enriched branched isomer content in humans. One hypothesis is that preferential biotransformation of branched PFOS precursors (e.g., perfluorooctane sulfonamides) results in an enrichment of branched PFOS. Isomer-specific biotransformation was recently investigated using mixture incubations of various

Table 9 PFOS isomer composition (%) in environmental and wildlife samples. Values shown are means unless stated otherwise

References	Sample	n-	iso-	5m/4m/3m	1m	tb + dimethyls	Other branched	Total branched	Analysis/comment
Houde et al. (2008)	PFOS standard (unknown supplier)	76.9	10.6	5.1	3.9	3.6		23.2	LC–MS/MS (PFP column, m/z 80 or 99). Quantification using n-PFOS standard (Wellington) adjusted using branched isomer response factors from Riddell et al. (2009)
	Water (range, Lake Ontario, 2004)	43–56	22–28	17–21	3.9–8.0	1.0–2.0		44–57	
	Sediment (range, Lake Ontario, 1995–2002)	81–89	4.6–10	2.2–5.9	2.7–4.4			11–19	
	Zooplankton (range, Lake Ontario, 2004 and 2006)	95–100	0.1–1.0	0.4–3.0	0			0–5	
	Mysis (Mysis relicta) (range, Lake Ontario, 2001)	91–92	3.9–5.2	1.7–2.4	0			8–9	
	Diporeia (Diporeia hoyi) (range, Lake Ontario, 2002 and 2003)	95–96	2.0–2.8	0.4–0.6	0.7–1.1			4–5	
	Alewife (Alosa pseudoharengus) (range, Lake Ontario, 2002)	90–91	4.2–7.1	2.3–2.4	2.5–2.8			8–9	
	Smelt (Osmerus mordax) (range, Lake Ontario, 2002)	88–92	4.6–4.8	1.8–2.3	0.5–0.9	0.4–0.5		8–12	
	Sculpin (Cottus cognatus) (range, Lake Ontario, 2002)	91–92	4.6–4.8	1.8–2.3	0.9–1.1	0.2–0.3		8–9	
	Lake Trout (Salvelinus namaycush) (range, Lake Ontario, 2002)	88–93	2.8–7.1	2.5–4.1	0.9–1.1	0.2–0.3		7–12	

Table 9 (continued)

References	Sample	n-	iso-	5m/4m/3m	1m	tb + dimethyls	Other branched	Total branched	Analysis/comment
Powley et al. (2008)	PFOS standard (Fluka)	74.0						26.0	LC–MS/MS (C8 column, m/z 80). Branched content determined using relative peak areas. Extent of isomer separation unclear
	Cod (Arctogadus glacialis) (Western Canadian Arctic, 2004)	50.0						50.0	
	Bearded and ringed Seal (Erignathus barbatus and Phoca hispida) (Western Canadian Arctic, 2004)	96.0						4.0	
Senthil Kumar et al. (2009)	Aquatic wildlife (GA, USA, 2006, 2007)	81 (range 77–89)						19 (11–23)	LC–MS/MS (C18 column, m/z 80). Branched content estimated from concentrations of linear/branched isomers in Senthil Kumar et al. (2009). Unclear how quantification of branched PFOS was conducted or the extent of isomer separation

Table 9 (continued)

References	Sample	n-	iso-	5m/4m/3m	1m	tb + dimethyls	Other branched	Total branched	Analysis/comment
Lloyd et al. (2009)	PFOS standard (Fluka)	~71						~29	LC–MS/MS (PFO column, m/z 80). Branched content estimated from peak height in chromatograms found in Lloyd et al. (2009). Branched isomers not baseline resolved
	Whitebait	~39						~61	
	Roe deer liver	~42						~58	
	Cromer crab	~86						~14	
	Carp roe	~89						~11	
Chu and Letcher (2009)	PFOS Standard (T-PFOS, Wellington)	65	11.3	19.7	0.9	1.9	1.1[a]	35	GC–MS (DB-5 column, ion monitored dependent on isomer). Branched content determined using isomer-specific quantification
	Herring gull egg (*Larus argentatus*) (Great Lakes, 1989)	94.5	2.9	2.2	0.2	0.1	0.1[a]	5.5	
	Double-crested cormorant egg (*Phalacrocorax auritus*) (Great Lakes, 2003)	95.9	2.2	1.4	0.3	0.1	0.2[a]	4.1	
	Polar bear plasma (*Ursus maritimus*) (Norwegian Arctic, 2007)	82.4	4.1	10.3	2.8	0.1	0.4[a]	17.6	
	Polar bear liver (*Ursus maritimus*) (Canadian Arctic, 2007, 2008)	92.4	4.1	2.8	0.3	0.1	0.4[a]	7.6	

[a] 2m-PFOS

concentrations of a technical PFOS precursor (NEtFOSA isomers) with cytochrome P450 isozymes (CYPs) 2C9, 2C19, and human liver microsomes (Benskin et al. 2009b). Isomer-specific biotransformation rate constants were significantly different at all concentrations, and the rank orders of these rate constants were different with two different isozyme systems. Furthermore, when the ECF mixture was incubated with human liver microsomes (containing all of the major CYP isozymes), isomer-specific biotransformation and product formation were also observed. These data cannot be extrapolated directly to predict the extent of isomer-specific PFOS accumulation from precursors in an environmental exposure scenario, whereby constant exposure, biotransformation, and elimination processes will all combine to achieve a steady state. Thus, further in vivo experiments are necessary with PFOS precursors. However, based on this early evidence it is reasonable to speculate that preferential biotransformation of branched PFOS-precursor isomers may result in enriched branched PFOS isomer patterns, thereby providing a possible explanation for the high abundance of branched PFOS isomers in some humans and wildlife (Tables 8 and 9) and a potential biomarker for exposure to precursors. This precursor hypothesis currently remains tentative, however, it could be confirmed by measuring non-racemic proportions of PFOS isomer enantiomers in biological samples, as described in a proof-of-principle study by Wang et al. (2009). In this work, a chiral, alpha-branched PFOS ($1m$-PFOS) precursor was observed to biotransform enantioselectively when incubated with human liver microsomes. Based on these results, PFOS source exposure in humans and wildlife may be determined by examination of enantiomeric fractions, although a method for separation of PFOS enantiomers requires development before this hypothesis can be tested.

PFOS isomer profiles in coastal Asia and the Atlantic Oceans were also recently examined and found to be very similar or slightly enriched in branched isomers compared to 3M ECF PFOS (Benskin et al. 2009c). On the contrary, enrichment of n-PFOS (i.e., >70% linear content), in comparison to 3M ECF PFOS, was never observed in ocean samples. Differential PFOS isomer pK_a values are unlikely to affect environmental partitioning since all PFOS isomers will be ionized at environmentally relevant pH, and it is unclear to what extent differential surface activity alone may influence boundary layer (water–air) partitioning of n-PFOS. Surface layer enrichment of n-PFOS could potentially result in water samples collected below the surface layer being enriched to some extent with branched isomers, and while this hypothesis remains tentative at this time, differences have been observed in total PFOS concentrations between surface microlayer and sub-surface water samples (Ju et al. 2008).

Other possible explanations for branched PFOS isomer enrichment in ocean water include degradation of ECF polymeric material containing unique isomer signatures or alternatively preferential abiotic degradation (Ochoa-Herrera et al. 2008; Yamamoto et al. 2007) of branched PFOS precursors. This may also explain the relative abundance of branched PFOS isomers in Lake Ontario (Houde et al. 2008), discussed above. In comparison, PFOS isomer profiles from coastal Asian locations (Shanghai, Tokyo Bay, Tomakomai Bay, and Japan Sea) were fairly consistent with 3M ECF PFOS for all samples (Benskin et al. 2009c).

6.3 Perfluorocarboxylate Isomer Profiles Other than PFOA

The source of branched long-chain carboxylates (C > 8) is still uncertain. Recent analysis by 3M and by this study (Table 3) supports the assertion by Prevedouros et al. (2006) that branched isomers of C4–C7 and C9–C13 may be present as residuals in both 3M ECF POSF-derived products and 3M ECF PFOA. Global emissions in 2000 have been estimated for C6–C13 PFCAs (Prevedouros et al. 2006), however, it is still uncertain whether PFCAs other than C4, C6, C8, and C9 were ever intentionally produced for large-scale manufacturing. Interestingly, recent data suggest the possibility of long-chain (i.e., C11, C13) sources of isopropyl-branched isomers (De Silva et al. 2009b; Furdui et al. 2008; Zushi et al. 2010). Unlike ECF which produces a variety of branched isomers, formation of isopropyl perfluoroalkyl compounds is possible via the telomerization reaction pathway of an isopropyl telogen.

De Silva et al. (2009b) detected no branched PFNA in precipitation, Lake Ontario sediment, or most Lake Ontario Biota (mysis, zooplankton, trout, alewife) but one branched isomer (*iso*-PFNA) in ringed seals from Resolute Bay and a single polar bear from the Canadian Arctic. Examination of Arctic lake sediment revealed four branched PFNA isomers, including *iso*-PFNA, along with *n*-PFNA. The surface water of this lake contained only *iso*-PFNA and *n*-PFNA. In Lake Ontario surface water and sediment, only *iso*-PFNA and *n*-PFNA were observed. In contrast, two branched isomers of PFNA have consistently been detected in isomer-specific monitoring of human blood (Benskin et al. 2006; De Silva et al. 2006, 2009b). Given the large number of patents describing the synthesis of isopropyl-branched PFCAs via telomerization (Katsushima et al. 1970; Millauer 1974; Katsushima et al. 1976; also see supporting info of De Silva et al. 2009b), it is realistic to expect that these compounds have experienced significant production. De Silva et al. (2009b) suggested that the presence of multiple branched isomers (i.e., isopropyl and monomethyls) was most likely suggestive of ECF inputs, but that detection of only the isopropyl isomer in the absence of other branched isomers was ambiguous with respect to ECF versus an isopropyl telomer source.

The hypothesis of intentional isopropyl PFCA production is supported by PFCA isomer profiles in archived Lake Trout (1979–2004) from Lake Ontario and archived suspended sediment from the Niagara River (1980–2003) (Furdui et al. 2008), as well as archived sediment cores from Tokyo Bay (from 1950s to 2004) (Zushi et al. 2010). Of the perfluorocarboxylates monitored in these studies, only PFUnA and PFTrA-branched isomers were detected consistently. In Tokyo Bay, a consistent increase in the ratio of branched to linear PFTrA isomers was observed from 1988 to 2004, suggesting increase in branched isomer production, while the opposite trend was observed in suspended sediment from the Niagara River, where the ratio of branched to linear isomers of PFUnA and PFTrA decreased significantly from 1980 to 2002. This latter trend was also observed in Lake Ontario trout, however, the rate of decrease of branched PFTrA in fish was statistically different than in sediment, in contrast to PFUnA, where trends in fish and sediment were consistent. The presence of branched isomers has typically been interpreted as an ECF contribution, however, the authors cite patents describing synthesis of isopropyl PFCAs by telomerization in both North America (Millauer 1974; Katsushima et al. 1970; cited in Furdui et al. 2008) and Japan (Katsushima et al. 1976; cited in Zushi et al. 2010) as evidence of isopropyl production sources in these regions. While the lack of branched PFOA in Lake Trout observed by Furdui et al. (2008) may be explained by the low bioaccumulation potential of most branched PFOA isomers (De Silva et al. 2009c), a subsequent study observed branched PFOA (two–four branched isomers) in all samples of Lake Ontario biota (2002, 2004, 2006), sediment (1998, 2002), and surface water (2001, 2002) (De Silva et al. 2009b). In addition, *iso*-PFNA was observed in surface water, sediment, and half of the biota samples (including trout). Interestingly, in this study isopropyl C9–C12 PFCAs were also observed, whereby PFUnA had notably higher branched isomer content (6–12%) than other long-chain PFCAs (<2%), consistent with the observation of the relatively high branched isomer content of PFUnA observed in Lake Trout (Furdui et al. 2008).

De Silva et al. (2009b) noted that Lake Ontario biota and dolphins from urban coastal areas in south-eastern United States and sporadic human blood samples contained an abundance of *iso*-PFUnA. Arctic samples, including Char Lake sediment, ringed seals, and polar bears had a different isomer profile in which *iso*-PFDoA (4–7% of total PFDoA) was dominant compared to *iso*-PFUnA (1–3%). In that study the authors speculated that atmospheric transport and oxidation of a precursor containing an isopropyl perfluoroundecyl moiety may be responsible. The same precursor could, presumably, also undergo biological transformation to yield *iso*-PFUnA, thus accounting for its presence at mid-latitudes that are heavily influenced by human activity. In coastal Asian waters a single branched isomer of PFNA, PFDA, PFUnA, and PFDoA was observed in addition to the respective linear isomer. There is currently a paucity of isomer-specific long-chain perfluorocarboxylate data to confirm whether these are consistent trends.

Benskin et al. (2007) examined perfluorocarboxylate (C9–C14) isomer profiles in pooled serum from pregnant Edmonton (Alberta, Canada) women and found, in addition to the linear isomer, two branched isomers of PFNA, and a single branched isomer of PFDA and PFUnA. In contrast, only linear isomers of PFDoA and PFTA were detectable. The peaks corresponding to branched isomers appeared to make up a relatively small component of the total concentrations based on relative LC–MS peak areas in several transitions. This is consistent with the observations by De Silva and Mabury (2006), who also detected two minor branched PFNA isomers and one minor branched PFUnA isomer in human serum, representing ∼1.6 and 2.3% of total PFNA and PFUnA concentrations, respectively.

7 Differences in Toxicity and Bioaccumulation of PFA Isomers

To date, most studies that examine biological properties of PFAs have not differentiated between the branched and linear structures. For PFOS, in vitro and in vivo experiments have typically relied on standards from Sigma–Aldrich/Fluka (∼20% branched) (Cui et al. 2009; Johansson et al. 2009), or 3M (∼30% branched) (Hu et al. 2002; Luebker et al. 2005; Seacat et al. 2003), representing a difference in branched isomer content of ~10%. For those studies in which the manufacturer is not identified, or the branched content unknown, it is reasonable to predict that the variability among studies is less than 15%, given the difference in branched isomer content in technical standards (Table 2), provided a linear standard is not used. Although isomer-specific toxicity information is lost when using technical standards, this is likely the most environmentally relevant choice of standards for PFOS (and its precursors), considering most exposure sources (water, dust, food, etc.), and the internal dose of organisms (from biomonitoring studies) show that multiple isomers are always present.

Biological testing of PFOA has also made use of standards predominantly from 3M (80% linear) or Sigma–Aldrich (~99% linear). It is unclear which standard is the most relevant for toxicity testing. For example, in humans and higher trophic level organisms, even though linear PFOA dominates the internal dose, this does not necessarily mean that co-exposure does not also occur to the multiple branched isomers via food or house dust (for humans). Nonetheless, the few toxicological comparisons of branched and linear isomers have suggested that there may be only subtle toxicological differences between linear and total branched isomers. Furthermore, it is unclear whether the differences are a function of reduced biological activity or less bioavailability of the branched isomers. For example, Loveless et al. (2006) compared the responses of rats and mice following exposure to either a technical, 77.6% n-PFOA/22.4% branched PFOA isomer mixture, enriched branched isomer dose of 54% $3m$-PFOA, 4% $4m$-PFOA, and 42% iso-PFOA or a 100% n-PFOA dose. Peroxisomal beta-oxidation was least pronounced after administration of the enriched branched dose, results that contrast somewhat to those of Vanden Heuvel et al. (2006), in which branched and linear PFOA were both able to activate peroxisome proliferator receptor (PPAR)-α to a similar peak effect in vitro. Body weights of rats and mice in the Loveless et al. (2006) study were also approximately 20% lower in rats, exposed to the mixture of linear and branched or pure linear doses, compared to the branched-only dose; this further supports the opinion that n-PFOA may be slightly more toxic in vivo. However, the authors also observed that n-PFOA was preferentially absorbed relative to $3m/4m$-PFOA and iso-PFOA at increasingly higher doses, suggesting that the increased potency of n-PFOA relative to branched, and branched + linear dosing regimens may simply be a result of decreased bioavailability of the branched isomers.

Other isomer-specific data in rodents and fish further corroborate the hypothesis that differences in toxicological response between branched and linear PFAs may be a result of differential bioavailability. For example, rats exposed to lower doses of PFHxS, PFOS, PFOA, and PFNA via a single gavage dose, or through a sub-chronic dietary exposure, showed varying degrees of selective retention of n-isomer, compared to the major branched isomers in ECF formulations (Benskin et al. 2009a; De Silva et al. 2009a). It should be noted that although most differences in excretion rates of PFOS isomers were not statistically significant, this could be reflective of the experimental design. For example, the single-dose exposure

(Benskin et al. 2009a) was likely not long enough to detect significant differences among most PFOS isomer rate constants. Likewise, in the sub-chronic exposure (De Silva et al. 2009a) PFOS isomer excretion rate constants were based only on a single animal. Despite these restrictions, some statistically significant differences were observed between n-PFOS compared to tb-PFOS and 4m-PFOS. It is likely that a longer depuration period along with a larger sample population would suggest preferential linear isomer retention compared to the major branched isomers in rodents, similar to observations by Sharpe et al. (2010) in fish. Rainbow trout exposed to PFOA and PFNA isomers through the diet showed a similar result, whereby the n-isomer was selectively retained in blood and tissues, relative to the majority of branched isomers (De Silva et al. 2009c). Similarly, Sharpe et al. (2010) demonstrated significant preferential accumulation of n-PFOS relative to branched isomers in rainbow trout and zebrafish. Also consistent with these results was the observation of a substantial enrichment of n-PFOS in Lake Ontario trout compared to water (Houde et al. 2008). Bioaccumulation factors calculated for n-PFOS in this study were estimated to be 3.4×10^4 L/kg, compared with 2.9×10^3 L/kg for the monomethyl-substituted isomers. The apparent difference in calculated values was attributed to enrichment of branched isomers in Lake Ontario water, however, it is not clear what is mediating this phenomenon. Possible explanations include preferential removal of n-PFOS to sediment or aerosols, albeit these hypotheses require further validation. Trophic magnification factors calculated for n-PFOS (4.6 ± 1.0), monomethyl-branched isomers (1.3 ± 0.17–2.6 ± 0.51), and dimethyl-branched isomers (no trophic magnification) also suggest that n-PFOS may preferentially biomagnify through the food chain, relative to branched isomers. Based on these results, it appears that exposure to any mixture of PFHxS, PFOS, PFOA, or PFNA isomers will result in enriched linear isomer profiles in a range of organisms.

Despite this general trend, there were some notable exceptions in which branched isomers were eliminated more slowly than the n-isomer of either PFOS or PFOA. The structure of the biopersistent PFOA isomers has not yet been determined, but the alpha-branch PFOS isomer (1m-PFOS) showed a remarkably long half-life (longer than the linear isomer) in male rats following single or sub-chronic dosing and was not significantly eliminated in female rats following sub-chronic dietary exposure. Interestingly, plots of half-life of linear and monomethyl-branched PFOS and PFOA isomers, in various species and dosing regimens, revealed a consistent structure–property relationship, whereby a relative decrease in pharmacokinetic half-life was observed as the branching point was moved from the perfluoroalkyl chain terminus (n-) to the 4m-position, whereas an increase in half-life was observed as the branching point moved from the 4m-position closer to the sulfonate or carboxylate group (Fig. 6).

While preferential elimination of branched PFHxS, PFOS, PFOA, and PFNA isomers reportedly occurs via the urine (Benskin et al. 2009a), it is still not clear what mechanism mediates this isomer-specific phenomenon. The sex hormone-mediated organic anion transporter (OAT) system plays an important role in the renal elimination of n-PFOA from male and female rats (Kudo et al. 2002). Katakura et al. (2007)

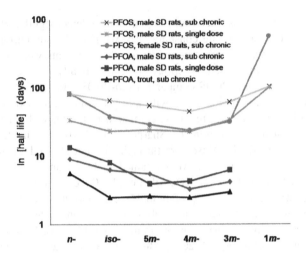

Fig. 6 Structure–property relationship between PFOS/PFOA isomers observed in rats (Benskin et al. 2009a; De Silva et al. 2009a) and fish (De Silva et al. 2009c). A relative decrease in half-life was observed as the branching point was moved from the perfluoroalkyl-chain terminus (n-) to the $4m$-position. In contrast, an increase in half-life was observed as the branching point moved from the $4m$-position closer to the sulfonate or carboxylate group regardless of sex, species, or dosing regimen. Note that while structure–property relationships between PFOS and PFOA isomers appear to be similar, half-lives for the individual branched PFOS isomers shown here were not statistically different from that of the n-isomer in either study

recently identified some of the specific transporters (oatp1 and OAT3) controlling the elimination of n-PFOA in rats, but it is still not clear if substrate–transporter binding is equivalent among isomers. In addition, while OAT3 was identified as mediating tubular uptake of n-PFOA and oatp1 with tubular reabsorption, the transporter responsible for tubular excretion of this PFA has not yet been identified. Sex differences in PFOA elimination in orally dosed fish have also been observed and are thought to be attributable to differences in renal transport activity (Lee and Schultz 2010). Clearly, there are many possibilities, and some transporters may have unique interactions with specific isomers. Only by probing the individual uptake of specific isomers by specific transporters can we obtain an accurate picture of the mechanism of elimination. It may be possible to correlate the preferential elimination of a given isomer in vivo to its affinity for renal uptake/excretion transporters and/or a lack of affinity for tubular reabsorption transporters.

Isomer-specific protein binding may also explain differential elimination rates in vivo. Branched PFAs with lower affinity for serum proteins could potentially undergo renal elimination to a greater extent than the linear isomer. Although it is known that n-PFOS and n-PFOA bind strongly to serum proteins (Jones et al. 2003; Ohmori et al. 2003), little is known about the binding affinity of branched isomers. Previous in vitro cytochrome P450 assays (Benskin et al. 2009b) are suggestive that differential protein binding can occur during the metabolism of PFA-precursor isomers, thus there is reason to suggest this could occur for PFA isomers and serum proteins as well. Such experiments would complement OATP and in vivo pharmacokinetic studies to get a better overall picture of the mechanism(s) of isomer-specific biological handling.

8 Summary

The two major manufacturing techniques for perfluorochemicals can be distinguished based on the isomeric profile of their products. ECF (major use from 1950s to 2002) results in a product containing both linear and branched isomers, while telomerization (major use from 2002 to present) typically yields an isomerically pure, linear product. Among the most important questions today, which has implications for future regulation of these chemicals, is to what extent human and environmental exposure is from historical products (i.e., ECF) versus currently manufactured fluorochemicals (i.e., telomer). Perfluoroalkyl-chain branching can also affect the physical and chemical properties of these chemicals, which may influence their environmental transport and degradation, partitioning, bioaccumulation, pharmacokinetics, and toxicity. Unless perfluorinated substances are considered as individual isomers, much of this information will be overlooked or missed altogether, which could potentially lead to inaccuracies in human and environmental risk assessments.

In this review, we have highlighted novel findings, current knowledge gaps, and areas for improvement based on early experiments on the disposition of PFA and PFA-precursor isomers in the environment. We have also emphasized the wealth of information that can potentially be gleaned from future work in this area, which renders routine adoption of isomer-specific methodologies an attractive and logical next step in the progression of fluorochemical analysis. However, despite vast improvements in recent years, a fast and comprehensive method capable of separating all major PFA and PFA-precursor isomers, while removing interferences is still required before these methods become routine in most labs. Purified and characterized standards of PFOA and PFOS that have isomer profiles consistent with those of historically produced (i.e., 3M) PFOS and PFOA are also required. The limited data available on PFA isomer profiles that exist in the environment and the biological properties of each isomer suggest that examination of isomer profiles may yield clues on the source of PFA contamination to humans and the environment. For example, contributions from historical versus current PFOA emissions can be quantified by examining the isomer profile in abiotic samples. Similarly, residual PFOS/PFOA in pre-2002 consumer products may be distinguished from directly emitted PFOS/PFOA by the existence of slight differences in isomer profile. PFOS signatures may also have the potential to distinguish between indirect exposure (via precursors) versus direct exposure (via the sulfonate), based on findings of isomer-specific and/or enantiospecific biotransformation in vitro. Isomer-specific monitoring extended to longer-chain PFAs may also be informative in determining current and historical exposure sources. Finally, given the recent increase of production of PFOSF-based chemicals, following their 2002 phase out, the ability of using isomer profiles to distinguish between historical and currently produced PFOS may also be possible.

Acknowledgments The 3M Co. is thanked for providing a generous donation of ECF FOSA, PFOS, and PFOA. Funding for NSERC VF (ADS) was provided by Chemicals Management Plan, Science and Risk Assessment Directorate to Derek Muir, EC Burlington. Funding for materials, supplies, and instrument time was provided through an NSERC Discovery Grant, an Alberta Ingenuity New Faculty Grant (JWM), and an Alberta Ingenuity graduate student scholarship (JPB). Alberta Health and Wellness is thanked for support of laboratory activities.

References

3M Co. (1999) U.S. Environmental Protection Agency Public Docket AR226-0550: fluorochemical use, distribution and release overview. U.S. Environmental Protection Agency, Office of Pollution Prevention and Toxic Substances, Washington, DC

Ahrens L, Felizeter S, Sturm R, Zie Z, Ebinghaus R (2009) Polyfluorinated compounds in waste water treatment plant effluents and surface waters along the River Elbe, Germany. Mar Pollut Bull 58:1326–1333

Apelberg BJ, Witter FR, Herbstman JB, Calafat AM, Halden RU, Needham LL, Goldman LR (2007) Cord serum concentrations of perfluorooctane sulfonate (PFOS) and perfluorooctanoate (PFOA) in relation to weight and size at birth. Environ Health Perspect 115:1670–1676

Appel AG, Abd-Elghafar SF (1990) Toxicity, sublethal effects, and performance of sulfluramid against the German cockroach (Dictyoptera: Blattellidae). J Econ Entomol 83:1409–1414

Armitage JM, Cousins IT, Buck RC, Prevedouros K, Russell MH, MacLeod M, Korzeniowski SH (2006) Modeling global-scale fate and transport of perfluorooctanoate emitted from direct sources. Environ Sci Technol 40:6969–6975

Armitage JM, MacLeod M, Cousins IT (2009a) Modeling the global fate and transport of perfluorooctanoic acid (PFOA) and perfluorooctanoate (PFO) emitted from direct sources using a multispecies mass balance model. Environ Sci Technol 43:1134–1140

Armitage JM, MacLeod M, Cousins IT (2009b) Comparative assessment of the global fate and transport pathways of long-chain perfluorocarboxylic acids (PFCAs) emitted from direct sources. Environ Sci Technol 43:5830–5836

Arsenault G, Chittim B, Gu J, McAlees A, McCrindle R, Robertson V (2008a) Separation and fluorine nuclear magnetic resonance spectroscopic (^{19}F NMR) analysis of individual branched isomers present in technical perfluorooctanesulfonic acid (PFOS). Chemosphere 73:S53–S59

Arsenault G, Chittim B, McAlees A, McCrindle R, Riddell N, Yeo B (2008b) Some issues relating to the use of perfluorooctanesulfonate (PFOS) samples as reference standards. Chemosphere 70:616–625

Balague J, Ameduri B, Boutevin B, Caporiccio G (1995) Synthesis of fluorinated telomers. Part 1: Telomerization of vinylidene fluoride with perfluoroalkyl iodides. J Fluorine Chem 70: 215–223

Bastosa JK, Freitasa LAP, Pagliarussia RS, Merinoc RE (2001) A rapid quantitative method for the analysis of sulfluramid and its isomers in ant bait by capillary column gas chromatography. J Sep Sci 24:406–410

Beesoon S, Webster G, Shoeib M, Harner T, Benskin JP, Martin JW (2009) Isomer profiles of perfluorinated compounds in house dust, pregnant women and cord serum: sources and transplacental transfer. Society of Environmental Toxicology and Chemistry 30th North American meeting, New Orleans, LA, USA, 19–23 Nov

Benskin JP, Bataineh M, Martin JW (2007) Simultaneous characterization of perfluoroalkyl carboxylate, sulfonate, and sulfonamide isomers by liquid chromatography–tandem mass spectrometry. Anal Chem 79:6455–6464

Benskin JP, De Silva AO, Martin LJ, Arsenault G, McCrindle R, Riddell N, Mabury SA, Martin JW (2009a) Disposition of perfluorinated acid isomers in Sprague Dawley rats. Part 1: Single dose. Environ Toxicol Chem 28:542–554

Benskin JP, Holt A, Martin JW (2009b) Isomer-specific biotransformation of a perfluorooctane sulfonate (PFOS)-precursor by cytochrome P450 isozymes and human liver microsomes. Environ Sci Technol 43:8566–8572

Benskin JP, Arhens LWY, Yamashita N, Taniyasu S, Lam PKS, Tomy G, Muir DC, Scott B, Spencer C, Rosenberg B, Martin JW (2009c) Perfluorinated acid isomer profiles in ocean water. Society of Environmental Toxicology and Chemistry 30th North American meeting, New Orleans, LA, USA, 19–23 Nov

Bernett MK, Zisman WA (1967) Surface properties of perfluoro acids as affected by terminal branching and chlorine substitution. J Phys Chem 71:2075–2082

Brace NO (1962) Long chain alkanoic and alkenoic acids with perfluoroalkyl terminal segments. J Org Chem 27:4491–4498

Burns DC, Ellis DA, Li H, McMurdo CJ, Webster E (2008) Experimental pK_a determination for perfluorooctanoic acid (PFOA) and the potential impact of pK_a concentration dependence on laboratory-measured partitioning phenomena and environmental modeling. Environ Sci Technol 42:9283–9288

Butenhoff J, Costa G, Elcombe C, Farrar D, Hansen K, Iwai H, Jung R, Kennedy G, Lieder P, Olsen G, Thomford P (2002) Toxicity of ammonium perfluorooctanoate in male cynomolgus monkeys after oral dosing for 6 months. Toxicol Sci 69:244–257

Butt CM, Muir DC, Stirling I, Kwan M, Mabury SA (2007) Rapid response of Arctic ringed seals to changes in perfluoroalkyl production. Environ Sci Technol 41:42–49

Chan E, Sandhu S, Benskin JP, Ralitsch M, Thibault N, Birkholz D, Martin JW (2009) Endogenous high-performance liquid chromatography/tandem mass spectrometry interferences and the case of perfluorohexane sulfonate (PFHxS) in human serum: Are we overestimating exposure? Rapid Commun Mass Spectrom 23:1405–1410

Cheng J, Psillakis E, Hoffmann MR, Colussi AJ (2009) Acid dissociation versus molecular association of perfluoroalkyl oxoacids: environmental implications. J Phys Chem A 113: 8152–8156

Chu S, Letcher RJ (2009) Linear and branched perfluorooctane sulfonate isomers in technical product and environmental samples by in-port derivatization-gas chromatography–mass spectrometry. Anal Chem 81:4256–4262

Cui L, Zhou QF, Liao CY, Fu JJ, Jiang GB (2009) Studies on the toxicological effects of PFOA and PFOS on rats using histological observation and chemical analysis. Arch Environ Contam Toxicol 56:338–349

De Silva AO, Benskin JP, Martin LJ, Arsenault G, McCrindle R, Riddell N, Martin JW, Mabury SA (2009a) Disposition of perfluorinated acid isomers in Sprague-Dawley rats. Part 2: Subchronic dose. Environ Toxicol Chem 28:555–567

De Silva AO, Mabury SA (2006) Isomer distribution of perfluorocarboxylates in human blood: potential correlation to source. Environ Sci Technol 40:2903–2909

De Silva AO, Mabury SA (2004) Isolating isomers of perfluorocarboxylates in polar bears (Ursus maritimus) from two geographical locations. Environ Sci Technol 38:6538–6545

De Silva AO, Muir DC, Mabury SA (2009b) Distribution of perfluorinated carboxylate isomers in select samples in the North American environment. Environ Toxicol Chem 28: 1801–1814

De Silva AO, Stock NL, Bonin J, Wong GW, Young C, Mabury SA (2008) Water solubility and octanol–water partition coefficient of perfluorooctylsulfonamides and fluorotelomer alcohols in: perfluorocarboxylate isomer analysis as a tool for source elucidation. PhD Thesis, Department of Chemistry, University of Toronto, ON, Canada

De Silva AO, Tseng PJ, Mabury SA (2009c) Toxicokinetics of perfluorocarboxylate isomers in rainbow trout. Environ Toxicol Chem 28:330–337

D'eon JC, Hurley M, Wallington TJ, Mabury SA (2006) Atmospheric chemistry of N-methyl per-fluorobutane sulfonamidoethanol, C$_4$F$_9$SO$_2$N(CH$_3$)CH$_2$CH$_2$OH: kinetics and mechanism of reaction with OH. Environ Sci Technol 40:1862–1868

Dinglasan-Panlilio MJ, Mabury SA (2006) Significant residual fluorinated alcohols present in various fluorinated materials. Environ Sci Technol 40:1447–1453

Ellis DA, Mabury SA, Martin JW, Muir DC (2001) Thermolysis of fluoropolymers as a potential source of halogenated organic acids in the environment. Nature 412:321–324

Ellis DA, Martin JW, De Silva AO, Mabury SA, Hurley MD, Andersen MPS, Wallington TJ (2004) Degradation of fluorotelomer alcohols: a likely atmospheric source of perfluorinated carboxylic acids. Environ Sci Technol 38:3316–3321

Ellis DA, Martin JW, Mabury SA, Hurley MD, Andersen MP, Wallington TJ (2003a) Atmospheric lifetime of fluorotelomer alcohols. Environ Sci Technol 37:3816–3820

Ellis DA, Martin JW, Muir DC, Mabury SA (2003b) The use of ^{19}F NMR and mass spectrometry for the elucidation of novel fluorinated acids and atmospheric fluoroacid precursors evolved in the thermolysis of fluoropolymers. Analyst 128:756–764

Ellis DA, Webster E (2009) Response to comment on: aerosol enrichment of the surfactant PFO and mediation of the water–air transport of gaseous PFOA. Environ Sci Technol 43:1234–1235

Fei C, McLaughlin JK, Lipworth L, Olsen J (2008) Prenatal exposure to perfluorooctanoate (PFOA) and perfluorooctanesulfonate (PFOS) and maternally reported developmental mile-stones in infancy. Environ Health Perspect 116:1391–1395

Fluoropolymer Manufacturing Group (2002) Fluoropolymer manufacturers group presentation slides. U.S. EPA Administrative Record AR226-1094

Furdui VI, Helm PA, Crozier PW, Lucaciu C, Reiner EI, Marvin CH, Whittle DM, Mabury SA, Tomy GT (2008) Temporal trends of perfluoroalkyl compounds with isomer analysis in Lake Trout from Lake Ontario (1979–2004). Environ Sci Technol 42:4739–4744

Gauthier SA (2004) Aqueous photolysis of the 8:2 fluorotelomer alcohol and the determina-tion of the water solubility of N-ethylperfluorooctanesulfoneamidoethyl alcohol. MSc Thesis, Department of Chemistry, University of Toronto, ON, Canada

Goss KU (2008a) Correction: the pK_a values of PFOA and other highly fluorinated carboxylic acids. Environ Sci Technol 42:5032

Goss KU (2008b) The pK_a values of PFOA and other highly fluorinated carboxylic acids. Environ Sci Technol 42:456–458

Grottenmuller R, Knaup W, Probst A, Dullinger B (2003) Process for the preparation of perfluorocarboxylic acids. US Patent 6,515,172 B2

Haszeldine RN (1953) Reactions of fluorocarbon radicals. Part 12. The synthesis of fluorocarbons and of fully fluorinated iodo-, bromo-, and chloroalkanes. J Chem Soc 3761–3768

Haug LS, Thomsen C, Becher G (2009) Time trends and the influence of age and gender on serum concentrations of perfluorinated compounds in archived human samples. Environ Sci Technol 43:2131–2136

Holzer J, Midasch O, Rauchfuss K, Kraft M, Reupert R, Angerer J, Kleeschulte P, Marschall N, Wilhelm M (2008) Biomonitoring of perfluorinated compounds in children and adults exposed to perfluorooctanoate-contaminated drinking water. Environ Health Perspect 116:651–657

Houde M, Czub G, Small JM, Backus S, Wang X, Alaee M, Muir DC (2008) Fractionation and bioaccumulation of perfluorooctane sulfonate (PFOS) isomers in a Lake Ontario food web. Environ Sci Technol 42:9397–9403

Houde M, Martin JW, Letcher RJ, Solomon KR, Muir DC (2006) Biological monitoring of polyfluoroalkyl substances: a review. Environ Sci Technol 40:3463–3473

Hu W, Jones PD, Upham BL, Trosko JE, Lau C, Giesy JP (2002) Inhibition of gap junctional inter-cellular communication by perfluorinated compounds in rat liver and dolphin kidney epithelial cell lines in vitro and Sprague-Dawley rats in vivo. Toxicol Sci 68:429–436

Johansson N, Eriksson P, Viberg H (2009) Neonatal exposure to PFOS and PFOA in mice results in changes in proteins which are important for neuronal growth and synaptogenesis in the developing brain. Toxicol Sci 108:412–418

Jones PD, Hu W, De Coen W, Newsted JL, Giesy JP (2003) Binding of perfluorinated fatty acids to serum proteins. Environ Toxicol Chem 22:2639–2649

Ju X, Jin Y, Sasaki K, Saito N (2008) Perfluorinated surfactants in surface, subsurface water and microlayer from Dalian Coastal waters in China. Environ Sci Technol 42:3538–3542

Karrman A, Langlois I, van Bavel B, Lindstrom G, Oehme M (2007) Identification and pattern of perfluorooctane sulfonate (PFOS) isomers in human serum and plasma. Environ Int 33: 782–788

Katakura M, Kudo N, Tsuda T, Hibino Y, Mitsumoto A, Kawashima Y (2007) Rat organic anion transporter 3 and organic anion transporting polypeptide 1 mediate perfluorooctanoic acid transport. J Health Sci 53:77–83

Katsushima A, Hisamoto I, Nagai M (1970) US Patent 3,525,758

Katsushima A, Hisamoto I, Nagai M, Fukui T, Kato T (1976) Method for water and oil repellent treatment. Japanese Patent 0831272 (in Japanese)

Kestner T (3M Co.) (1997) U.S. Environmental Protection Agency Public Docket AR226-0564: fluorochemical isomer distribution by ^{19}F-NMR spectroscopy. U.S. Environmental Protection Agency, Office of Pollution Prevention and Toxic Substances, Washington, DC

Kissa E (2005) Fluorinated surfactants and repellents, 2nd edn. Marcel Dekker, NewYork

Korkowski P, Kestner T (1999) Chemical characterization of FOSA & PFOSA samples by ^{1}H and ^{19}F NMR spectroscopy. Provided by 3M Co, 21 Jan 2009

Kudo N, Katakura M, Sato Y, Kawashima Y (2002) Sex hormone-regulated renal transport of perfluorooctanoic acid. Chem Biol Interact 139:301–316

Kutsuna S, Hori J (2008) Experimental determination of Henry's law constant of perfluorooctanoic acid (PFOA) at 298 K by means of an inert-gas stripping method with a helical plate. Atmos Environ 42:8883–8892

Langlois I (2006) Mass spectrometric isomer characterization of perfluorinated compounds in technical mixture, water, and human blood. PhD Thesis, Department of Chemistry, University of Basel, Switzerland

Langlois I, Berger U, Zencak Z, Oehme M (2007) Mass spectral studies of perfluorooctane sulfonate derivatives separated by high-resolution gas chromatography. Rapid Commun Mass Spectrom 21:3547–3553

Langlois I, Oehme M (2006) Structural identification of isomers present in technical perfluorooctane sulfonate by tandem mass spectrometry. Rapid Commun Mass Spectrom 20:844–850

Lau C, Anitole K, Hodes C, Lai D, Pfahles-Hutchens A, Seed J (2007) Perfluoroalkyl acids: a review of monitoring and toxicological findings. Toxicol Sci 99:366–394

Lee JJ, Schultz IR (2010) Sex differences in the uptake and disposition of perfluorooctanoic acid in fathead minnows after oral dosing. Environ Sci Technol 44:491–497

Lehmler HJ (2005) Synthesis of environmentally relevant fluorinated surfactants – a review. Chemosphere 58:1471–1496

Lieder PH, Chang SC, York RG, Butenhoff JL (2009) Toxicological evaluation of potassium perfluorobutanesulfonate in a 90-day oral gavage study with Sprague-Dawley rats. Toxicology 255:45–52

Lloyd AS, Bailey VA, Hird SJ, Routledge A, Clarke DB (2009) Mass spectral studies towards more reliable measurement of perfluorooctanesulfonic acid and other perfluorinated chemicals (PFCs) in food matrices using liquid chromatography/tandem mass spectrometry. Rapid Commun Mass Spectrom 23:2923–2938

Lopez-Fontan JL, Sarmiento F, Schulz PC (2005) The aggregation of sodium perfluorooctanoate in water. Colloid Polym Sci 283:862–871

Loveless SE, Finlay C, Everds NE, Frame SR, Gillies PJ, O'Connor JC, Powley CR, Kennedy GL (2006) Comparative responses of rats and mice exposed to linear/branched, linear, or branched ammonium perfluorooctanoate (APFO). Toxicol 220:203–217

Luebker DJ, York RG, Hansen KJ, Moore JA, Butenhoff JL (2005) Neonatal mortality from in utero exposure to perfluorooctanesulfonate (PFOS) in Sprague-Dawley rats: dose–response, and biochemical and pharamacokinetic parameters. Toxicology 215:149–169

Lv G, Wang L, Liu S, Li S (2009) Determination of perfluorinated compounds in packaging materials and textiles using pressurized liquid extraction with gas chromatography–mass spectrometry. Anal Sci 25:425–429

Martin JW, Ellis DA, Mabury SA, Hurley MD, Wallington TJ (2006) Atmospheric chemistry of perfluoroalkanesulfonamides: kinetic and product studies of the OH radical and Cl atom initiated oxidation of N-ethyl perfluorobutanesulfonamide. Environ Sci Technol 40: 864–872

Martin JW, Kannan K, Berger U, de Voogt P, Field J, Franklin J, Giesy JP, Harner T, Muir DC, Scott B, Kaiser M, Jarnberg U, Jones KC, Mabury SA, Schroeder H, Simcik M, Sottani C, van Bavel B, Karrman A, Lindstrom G, van Leeuwen SP (2004) Analytical challenges hamper perfluoroalkyl research. Environ Sci Technol 38:248A–255A

Martin JW, Mabury SA, O'Brien PJ (2005) Metabolic products and pathways of fluorotelomer alcohols in isolated rat hepatocytes. Chem Biol Interact 155:165–180

McMurdo CJ, Ellis DA, Webster E, Butler J, Christensen RD, Reid LK (2008) Aerosol enrichment of the surfactant PFO and mediation of the water–air transport of gaseous PFOA. Environ Sci Technol 42:3969–3974

Millauer H (1974) US Patent 3,829,512

Ochoa-Herrera V, Sierra-Alvarez R, Somogyi A, Jacobsen NE, Wysocki VH, Field JA (2008) Reductive defluorination of perfluorooctane sulfonate. Environ Sci Technol 42:3260–3264

OECD (2002) OECD document ENV/JM/RD(2002)17/FINAL: hazard assessment of PFOS. Organisation for Economic Co-operation and Development Environment Directorate, Paris

Ohmori K, Kudo N, Katayama K, Kawashima Y (2003) Comparison of the toxicokinetics between perfluorocarboxylic acids with different carbon chain length. Toxicology 184:135–140

Olsen GW, Burris JM, Ehresman DJ, Froehlich JW, Seacat AM, Butenhoff JL, Zobel LR (2007) Half-life of serum elimination of perfluorooctanesulfonate, perfluorohexanesulfonate, and perfluorooctanoate in retired fluorochemical production workers. Environ Health Perspect 115:1298–1305

Olsen GW, Chang SC, Noker PE, Gorman GS, Ehresman DJ, Lieder PH, Butenhoff JL (2009) A comparison of the pharmacokinetics of perfluorobutanesulfonate (PFBS) in rats, monkeys, and humans. Toxicology 256:65–74

Parsons JR, Sáez M, Dolfing J, de Voogt P (2008) Biodegradation of perfluorinated compounds. Rev Environ Contam Toxicol 196:53–71

Paul AG, Jones KC, Sweetman AJ (2009) A first global production, emission, and environmental inventory for perfluorooctane sulfonate. Environ Sci Technol 43:386–392

Perrin DD, Dempsey B, Serjeant EP (1981) pK_a prediction for organic acids and bases. Chapman and Hall, London

Plumlee MH, McNeill K, Reinhard M (2009) Indirect photolysis of perfluorochemicals: hydroxyl radical-initiated oxidation of N-ethyl perfluorooctane sulfonamido acetate (NEtFOSAA) and other perfluoroalkane sulfonamides. Environ Sci Technol 43:3662–3668

Powley CR, George SW, Russell MH, Hoke RA, Buck RC (2008) Polyfluorinated chemicals in a spatially and temporally integrated food web in the Western Arctic. Chemosphere 70: 664–672

Prevedouros K, Cousins IT, Buck RC, Korzeniowski SH (2006) Sources, fate and transport of perfluorocarboxylates. Environ Sci Technol 40:32–44

Rayne S, Forest K, Friesen KJ (2009) Computational approaches may underestimate pK(a) values of longer-chain perfluorinated carboxylic acids: implications for assessing environmental and biological effects. J Environ Sci Health A Tox Hazard Subst Environ Eng 44: 317–326

Rayne S, Forest K, Friesen KJ (2008a) Relative gas-phase free energies for the C3 through C8 linear and branched perfluorinated sulfonic acids: implications for kinetic versus thermodynamic control during synthesis of technical mixtures and predicting congener profile inputs to environmental systems. J Mol Struct: Theochem 869:81–83

Rayne S, Forest K, Friesen KJ (2008b) Congener-specific numbering systems for the environmentally relevant C4 through C8 perfluorinated homologue groups of alkyl sulfonates, carboxylates, telomer alcohols, olefins, and acids, and their derivatives. J Environ Sci Health A Tox Hazard Subst Environ Eng 43:1391–1401

Reagen WK, Lindstrom KR, Jacoby CB, Purcell RG, Kestner TA, Payfer RM, Miller JW (2007) Environmental characterization of 3M electrochemical fluorination derived perfluorooctanoate and perfluorooctanesulfonate. Society of Environmental Toxicology and Chemistry 28th North American meeting, Milwaukee, WI, USA, 11–15 Nov

Rhoads KR, Janssen EM, Luthy RG, Criddle CS (2008) Aerobic biotransformation and fate of N-ethyl perfluorooctane sulfonamidoethanol (N-EtFOSE) in activated sludge. Environ Sci Technol 42:2873–2879

Riddell N, Arsenault G, Benskin JP, Chittim B, Martin JW, McAlees A, McCrindle R (2009) Branched perfluorooctane sulfonate isomer quantification and characterization in real samples by HPLC/ESI-MS(/MS). Environ Sci Technol 43:7902–7908

Ruisheng Y (2008) Additional information of production and use of PFOS. Fax from Ministry of Environmental Protection of China. Stockholm Convention Secretatiat, Geneva, Switzerland. http://chm.pops.int/Portals/0/Repository/addinfo_2008/UNEP-POPS-POPRC-SUB-F08-PFOS-ADIN-CHI.English.pdf

Rylander C, Brustad M, Falk H, Sandanger TM (2009a) Dietary predictors and plasma concentrations of perfluorinated compounds in a coastal population from northern Norway. J Environ Pub Health 2009:1–10

Rylander C, Duong Trong P, Odland JO, Sandanger TM (2009b) Perfluorinated compounds in delivering women from south central Vietnam. J Environ Monit 11:2002–2008

Santoro MA (2009) Activities at 3M and Dyneon LLC to develop PFC alternatives. Workshop on managing perfluorinated chemicals and transitioning to safer alternatives, Geneva, Switzerland, February 12–13

Savu PM (1994) Fluorinated higher carboxylic acids. In: Kirk-Othmer encyclopedia of chemical technology, John Wiley & Sons, USA

Seacat AM, Thomford PJ, Hansen KJ, Clemen LA, Eldridge SR, Elcombe CR, Butenhoff JL (2003) Sub-chronic dietary toxicity of potassium perfluorooctanesulfonate in rats. Toxicology 183:117–131

Seacat AM, Thomford PJ, Hansen KJ, Olsen GW, Case MT, Butenhoff JL (2002) Subchronic toxicity studies on perfluorooctanesulfonate potassium salt in cynomolgus monkeys. Toxicol Sci 68:249–264

Senthil Kumar K, Zushi Y, Masunaga S, Gilligan M, Sajwan SS (2009) Perfluorinated organic contaminants in sediment and aquatic wildlife, including sharks, from Georgia, USA. Mar Pollut Bull 58:601–634

Sharpe RL, Benskin JP, Laarman AL, MacLeod SM, Martin JW, Wong CS, Goss GG (2010) Perfluorooctane sulfonate toxicity, isomer-specific accumulation, and maternal transfer in zebrafish (*Danio rerio*) and rainbow trout (*Oncorhynchus mykiss*). Environ Toxicol Chem (in press)

Shelton K (2009) DuPont approach to PFOA stewardship. Presentation at international workshop on managing perfluorinated chemicals and transitioning to safer alternatives, Geneva, Switzerland, 12–13 Feb

Shin-ya S (2009) PFC activities at Asahi. Presentation at international workshop on managing perfluorinated chemicals and transitioning to safer alternatives, Geneva, Switzerland, 12–13 Feb

Simons JH (1949) Electrochemical process for the production of fluorocarbons. J Electrochem Soc 95:47–59

Skutlarek D, Exner M, Färber H (2006) Perfluorinated surfactants in surface and drinking waters. Environ Sci Pollut Res 13:299–307

Smart BE (2001) Fluorine substituent effects (on bioactivity). J Fluorine Chem 109:3–8

Stevenson L (2002) U.S. Environmental Protection Agency Public Docket AR-2261150: com-
   parative analysis of fluorochemicals in human serum samples obtained commercially. U.S.
   Environmental Protection Agency, Office of Pollution Prevention and Toxic Substances,
   Washington, DC
Tomy GT, Tittlemier SA, Palace VP, Budakowski WR, Braekevelt E, Brinkworth L, Friesen
   K (2004) Biotransformation of N-ethyl perfluorooctanesulfonamide by rainbow trout
   (Oncorhynchus mykiss) liver microsomes. Environ Sci Technol 38:758–762
Vanden Heuvel JP, Thompson JT, Frame SR, Gillies PJ (2006) Differential activation of nuclear
   receptors by perfluorinated fatty acid analogs and natural fatty acids: a comparison of human,
   mouse, and rat peroxisome proliferator-activated receptor-alpha, -beta, and -gamma, liver X
   receptor-beta, and retinoid X receptor-alpha. Toxicol Sci 92:476–489
Vyas SM, Kania-Korwel I, Lehmler HJ (2007) Differences in the isomer composition of perfluorooctanesulfonyl (PFOS) derivatives. J Environ Sci Health A Tox Hazard Subst Environ Eng
   42:249–255
Wallington TJ, Hurley MD, Xia J, Wuebbles DJ, Sillman S, Ito A, Penner JE, Ellis DA, Martin J,
   Mabury SA, Nielsen OJ, Sulbaek Andersen MP (2006) Formation of C7F15COOH (PFOA) and
   other perfluorocarboxylic acids during the atmospheric oxidation of 8:2 fluorotelomer alcohol.
   Environ Sci Technol 40:924–930
Wang Y, Arsenault G, Riddell N, McCrindle R, McAlees A, Martin JW (2009) Perfluorooctane
   sulfonate (PFOS) precursors can be metabolized enantioselectively: principle for a new PFOS
   source tracking tool. Environ Sci Technol 43:8283–8289
Wania F (2007) A global mass balance analysis of the source of perfluorocarboxylic acids in the
   Arctic Ocean. Environ Sci Technol 41:4529–4535
Wellington Laboratories (2005) Certificate of analysis/documentation for T-PFOS. Obtained Mar
   2006 from Wellington Laboratories, Guelph, ON, Canada
Wellington Laboratories (2007) Certificate of analysis/documentation for br-PFOSK. Obtained Jan
   2008 from Wellington Laboratories, Guelph, ON, Canada
Wellington Laboratories (2008) Certificate of analysis/documentation for individual PFOS/PFOA
   isomers. Obtained Sept 2008 from Wellington Laboratories, Guelph, ON, Canada
Wendling L (2003) U.S. Environmental Protection Agency Public Docket OPPT-2003-0012-0007:
   environmental, health and safety measures relating to perfluorooctanoic acid and its salts
   (PFOA). U.S. Environmental Protection Agency, Office of Pollution Prevention and Toxic
   Substances, Washington, DC
White SS, Kato K, Jia LT, Basden BJ, Calafat AM, Hines EP, Stanko JP, Wolf CJ, Abbott BD,
   Fenton SE (2009) Effects of perfluorooctanoic acid on mouse mammary gland development and
   differentiation resulting from cross-foster and restricted gestational exposures. Reprod Toxicol
   29:289–298
Xu L, Krenitsky DM, Seacat AM, Butenhoff JL, Anders MW (2004) Biotransformation of N-ethyl-
   N-(2-hydroxyethyl)perfluorooctanesulfonamide by rat liver microsomes, cytosol, and slices and
   by expressed rat and human cytochromes P450. Chem Res Toxicol 17:767–775
Yamamoto T, Noma Y, Sakai S, Shibata Y (2007) Photodegradation of perfluorooctane sulfonate
   by UV irradiation in water and alkaline 2-propanol. Environ Sci Technol 41:5660–5665
Young CJ, Furdui VI, Franklin J, Koerner RM, Muir DC, Mabury SA (2007) Perfluorinated acids
   in Arctic snow: new evidence for atmospheric formation. Environ Sci Technol 41:3455–3461
Zushi Y, Tamada M, Kanai Y, Masunaga S (2010) Time trends of perfluorinated compounds from
   the sediment core of Tokyo Bay, Japan (1950s–2004). Environ Pollut 158:756–763

Biodegradation of Fluorinated Alkyl Substances

Tobias Frömel and Thomas P. Knepper

Contents

1 Introduction

The interest in the fate of fluorinated chemicals in the environment emerged in the 1980s–1990s, when chlorofluorocarbons were banned due to their extreme potential for destroying the ozone layer (Molina and Rowland 1974; Rowland 2002). More recent concerns connected with fluorinated chemicals have arisen within the last decade when the perfluorinated carboxylic acids (PFCAs) and perfluorinated alkane sulfonates (PFASs) were found to be environmentally ubiquitous (Giesy and Kannan 2001; Hansen et al. 2001). Indeed, the presence of these chemicals in human blood serum had been surmised in the 1960s (Taves 1968), but at that time, analytical methods were not yet capable of identifying these compounds.

Environmental contamination by these compounds soon entailed profound interdisciplinary studies to assess their hazards, both to the environment and to humans, as well as to determine their sources. As to sources, biodegradation has been proven to contribute significantly to the amounts of perfluorinated compounds (PFCs)

T.P. Knepper (✉)
Institute for Analytic Research, Hochschule Fresenius, 65510 Idstein, Germany
e-mail: knepper@hs-fresenius.de

P. de Voogt (ed.), *Reviews of Environmental Contamination and Toxicology*,
Reviews of Environmental Contamination and Toxicology 208,
DOI 10.1007/978-1-4419-6880-7_3, © Springer Science+Business Media, LLC 2010

detected globally, especially for perfluorooctanoic acid (PFOA). Prevedouros et al. (2006) estimated that the degradation of perfluorooctanesulfonyl fluoride-based and fluorotelomer-based chemicals contributes from 0.1 to 5% of the total PFC environmental burden. The contribution to perfluorooctane sulfonate (PFOS) contamination, however, is estimated to be only minor (Paul et al. 2008). A main purpose of performing biodegradation studies of fluorinated chemicals is to evaluate whether PFCA and PFAS are formed, because these compounds are known to exhibit adverse effects (Lau et al. 2007; Peden-Adams et al. 2008; Liao et al. 2009). The Organization of Economic Co-operation and Development (OECD) has published a list containing 615 chemicals that may degrade into PFCAs (Organization of Economic Co-operation and Development 2007). However, only a small number of these chemicals have actually been shown to contribute to environmental levels of the PFCAs.

Standardized biodegradation tests, such as those provided by the OECD, are mainly based on measurements of sum parameters, such as dissolved organic carbon (Organization of Economic Co-operation and Development 1992) or CO_2 evolution (US Environmental Protection Agency 2008); thus, no chemical-specific analyses are performed. Critical metabolites, i.e., those which may accumulate in the environment or pose a risk to the environment and humans, cannot be identified with the techniques used. Therefore, in recent years, numerous biodegradation studies have been performed in which superior analytical methods have been used, most of which are based on hyphenated mass spectrometry (MS). With the help of these methods, several novel metabolites have been identified and their levels in environmental compartments partly delineated.

In this review, we summarize the accomplishments that have been made in understanding the metabolism of fluorinated alkyl substances that have a range of fluorine content.

2 The Persistence of Perfluorinated Surfactants

By definition, a perfluorinated surfactant is one which does not contain any hydrogen atoms attached to carbon atoms. The most prominent perfluorinated surfactants are PFOS and PFOA (Table 1 shows chemical formulae and acronyms of the compounds addressed in this chapter). Their non-degradability under both aerobic and anaerobic conditions has been proven in several studies (Remde and Debus 1996; Schröder 2003; Sáez et al. 2008). Even under sulfur-limiting conditions, PFOS underwent no biodegradation (Key et al. 1998). In only one study it is claimed that PFOS and PFOA can be anaerobically degraded, although the results presented are highly doubtful and are more likely attributable to sorption phenomena, because no increase in fluoride concentration could be determined (Schröder 2003).

The persistence of such chemicals that contain long perfluorinated alkyl chains may be attributed to the very strong C–F bond ($\Delta H \approx 407$ kJ/mol). Besides, there have been only very few indications that naturally occurring defluorinating enzymes

Table 1 Names, acronyms, structures, sources, and applications of low molecular weight fluorinated alkyl substances (y = yes, n = no)

Acronym (trivial) name[a]	Structure[a]	Sources/main applications	(Potentially)
7:2-sFTOH (sec.-7:2-fluorotelomer alcohol)	$F\text{-}(CF_2)_7\text{-}CH(OH)\text{-}CH_3$	Metabolite of FTOH	No
7:3-FTA (7:3-fluorotelomer acid)	$F\text{-}(CF_2)_7\text{-}CH_2\text{-}CH_2\text{-}C(=O)OH$	Metabolite of FTOH	Yes
7:3-FTUA (7:3-fluorotelomer unsaturated acid)	$F\text{-}(CF_2)_7\text{-}CH=CH\text{-}C(=O)OH$	Metabolite of FTOH	No
8:2-FTA (8:2-fluorotelomer acid)	$F\text{-}(CF_2)_8\text{-}CH_2\text{-}C(=O)OH$	Metabolite of FTOH	No
8:2-FTAl (8:2-fluorotelomer aldehyde)	$F\text{-}(CF_2)_8\text{-}CH_2\text{-}C(=O)H$	Metabolite of FTOH	No
8:2-FTOH (8:2-fluorotelomer alcohol)	$F\text{-}(CF_2)_8\text{-}CH_2\text{-}CH_2\text{-}OH$	Intermediate in fluoropolymer manufacture	No
8:2-FTS (8:2-fluorotelomer sulfonate)	$F\text{-}(CF_2)_6\text{-}CH_2\text{-}CH_2\text{-}SO_3^-$	Aqueous fire-fighting foams	No
8:2-FTUA (8:2-fluorotelomer unsaturated acid)	$F\text{-}(CF_2)_7\text{-}CF=CH\text{-}C(=O)OH$	Metabolite of FTOH	No
BTFMA-AS (ω-[bis(trifluoromethyl)amino] alkane-1-sulfonates)	$(F_3C)_2N\text{-}(CH_2)_n\text{-}SO_3^-$, $n = 7\text{--}13$	None/model compound	No
DFMS (difluoromethane sulfonate)	$F_2HC\text{-}SO_3^-$	None/model compound	No
FOSA (perfluorooctanesulfonamide)	$F\text{-}(CF_2)_8\text{-}S(=O)_2\text{-}NH_2$	Metabolite of N-EtFOSE	No
FOSAA (perfluorooctanesulfonamido-acetate)	$F\text{-}(CF_2)_8\text{-}S(=O)_2\text{-}NH\text{-}CH_2\text{-}C(=O)OH$	Metabolite of N-EtFOSE	No
FTEO (fluorotelomer ethoxylate)	$F\text{-}(CF_2\text{-}CF_2)_x(CH_2\text{-}CH_2\text{-}O)_y\text{-}H$	Nonionic fluorosurfactant Additive in paints, coatings,...	No

Table 1 (continued)

Compound	Structure	Use/Source	
FTEOC (fluorotelomer ethoxycarboxylate)		Metabolite of FTEO	Yes[b]
N-EtFOSA (N-ethyl perfluorooctane sulfonamide)		Metabolite of N-EtFOSE	No
N-EtFOSAA (N-ethyl perfluorooctane sulfonamidoacetic acid)		Metabolite of N-EtFOSE	No
N-EtFOSE (N-ethyl perfluorooctane sulfonamidoethanol)		Intermediate in fluoropolymer manufacture	No
PFOA (perfluorooctanoic acid)		Additive in fluoropolymer manufacture Fluoropolymer dispersion agent	Yes
PFOS (perfluorooctane sulfonate)		Aqueous fire-fighting foams Metabolite of N-EtFOSE	Yes
PFOSI (perfluorooctane sulfinate)		Metabolite of N-EtFOSE	No
TFA (trifluoroacetic acid)		Widely used synthetic reagent Model compound	Yes
TFES (2,2,2-trifluoroethane sulfonate)		None/model compound	No
TFM-DS (10-trifluoromethoxy-decane-1-sulfonate)		None/model compound	No
TFMP-NS (9-[4-(trifluoromethyl)phenoxy]nonane-1-sulfonate)		None/model compound	No
TFMS (trifluoromethane sulfonate)		None/model compound	Yes

[a]Carboxylic acids are shown as the protonated free acids, although most of them are unlikely to occur in this state in the environment due to their high acidity. Published studies often differentiate between the free acid and the respective anion, which was not done here.
[b]FTEOCs with $y<9$ are not further degraded.

exist, which is probably a result of the scarcity of fluorinated molecules in nature and the high bond enthalpy. Indeed, only 4-fluorothreonine, several fluorinated fatty acids (as well as fluorocitrate), fluoroacetone, and one fluorinated nucleoside derivative (nucleocidin) have, so far, been discovered to naturally occur (O'Hagan and Harper 1999; Murphy et al. 2003). In mammalian species, cytochrome P_{450} monooxygenases can defluorinate certain molecules, but the number of published articles that address microbial defluorination is very scarce, at least for aliphatic compounds. Aromatic fluoroorganics seem to be more susceptible to defluorination, because the number of aromatic compounds reported to be defluorinated exceeds that of aliphatic compounds by far (Natarajan et al. 2005).

In a study performed under sulfur-limiting conditions, the degradability of PFOS, $1H,1H,2H,2H$-PFOS (8:2-fluorotelomersulfonate, 8:2-FTS), trifluoromethane sulfonate (TFMS), difluoromethane sulfonate (DFMS), and 2,2,2-trifluoroethane sulfonate (TFES) was investigated with respect to desulfonation and release of inorganic fluoride (Key et al. 1998). The substance 8:2-FTS is used in aqueous fire-fighting foams and has been proven to leach into groundwater after use (Schultz et al. 2004). It was clearly shown that only those compounds bearing at least one hydrogen atom in the α-position relative to the sulfonate group are desulfonated, namely, 8:2-FTS, DFMS, and TFES. Defluorination was only observed for 8:2-FTS and TFES. The metabolites generated remain unknown. Although sulfur-limiting conditions are not representative of standard environmental conditions, if desulfonation does not occur under these extreme conditions, it is unlikely to occur under more natural environmental conditions.

The thermodynamic aspect of defluorination has been very thoroughly assessed (Parsons et al. 2008). To summarize the foregoing, the implementation and environmental use of long perfluorinated alkyl chains is very likely to create non-biodegradable compounds despite the fact that defluorination is a thermodynamically favored reaction. If biotic transformations occur at all, they do not proceed to completion, because of the recalcitrance of the perfluorinated moiety in the molecule; therefore, long-lived highly fluorinated metabolites will be environmentally persistent.

3 Understanding the Complex Biodegradation of Fluorotelomer-based Chemicals

A number of highly fluorinated chemicals have been subjected to biodegradation studies. Among these, the fluorotelomer-based chemicals have recently attracted scientific attention. Such attention has resulted from both their complex metabolism and the high volume of their production. The most prominent fluorotelomer-based chemicals are fluorotelomer alcohols (FTOHs) and fluorotelomer olefins, which are mainly used as chemical intermediates in the production of fluorinated polymers.

Biodegradation of FTOHs has been studied in soil, wastewater, and mineral media amended with bacteria originating from different environmental compartments. One of the biggest challenges, when investigating FTOH degradation, is their relatively high volatility (boiling point – 8:2-FTOH $\approx$ 80°C), vapor pressure (8:2-FTOH: 3 Pa at 25°C; Liu and Lee 2005), and poor water solubility (137 $\pm$ 53 μg/L at 25°C and 194 $\pm$ 32 μg/L at 22.3 $\pm$ 0.4°C), which results in a relatively high water–air partitioning coefficient (Lei et al. 2004; Goss et al. 2006). Furthermore, when dissolved in water, the tendency of the FTOHs to sorb onto particles is very high (Wang et al. 2005a). Such sorption was investigated in detail by Liu et al. (2007), who found a log K_{OC} of 4.13 $\pm$ 0.16, but a log K_{DOC} value of 5.3 $\pm$ 0.29 and partially irreversible binding of 8:2-FTOH to dissolved organic carbon (DOC; Liu and Lee 2005).

The striking partitioning behavior of FTOHs complicates the understanding of the fate of these compounds in several ways: Firstly, mass balance is difficult to achieve in laboratory experiments, and this can only be remedied by working with closed systems, which in turn, do not represent natural environmental conditions. Secondly, owing to their high volatility, diverse atmospheric chemical reactions may occur, which may finally also lead to PFCAs. Although we shall not address this point in detail here, several papers have been published on this topic (Ellis et al. 2003, 2004; Wallington et al. 2006). Briefly, atmospheric chemical reactions of FTOHs with radicals may lead to the formation of different PFCAs; thus, these reactions must be taken into account when evaluating sources of PFCAs. Below, the microbial degradation of 8:2-FTOH is described in detail. It is probable that the degradation of other FTOHs is similar, although little effort has been put on these other chemicals, as yet.

In general, the biodegradation of 8:2-FTOH begins with the oxidation of the hydroxyl function to the respective aldehyde (8:2-FTAl) and to the carboxylic acid (8:2-FTA) (Fig. 1). The 8:2-FTA is then transformed to the unsaturated carboxylic acid (8:2-FTUA) by formal cleavage of hydrogen fluoride, representing the next step within the circle of β-oxidation. Both FTA and FTUA of various chain lengths have been found to occur at low concentrations in environmental samples (Furdui et al. 2007; Butt et al. 2007). It appears that 8:2-FTUA can be further metabolized by several pathways. One such pathway gives rise to the 7:3-FTUA and then to the potentially persistent 7:3-FTA, which was first described in the literature by Wang et al. (2005b). If this step can be confirmed, this would be one of the first reports of microbial defluorination (Key et al. 1997). Also perfluorohexanoic acid (PFHxA) is generated via 7:3-FTUA by unknown metabolic pathways. Although not detected in other studies, the PFHxA may occur at levels up to about 4 mol% (Wang et al. 2009).

Also important, from the research and environmental regulatory standpoint, is that the metabolic conversion of 8:2-FTOH leads to PFOA, which has been detected in all degradation studies of 8:2-FTOH performed so far. The elucidation of the metabolic pathway leading to PFOA is still a matter of ongoing research. While initial studies proposed that this metabolic pathway proceeded from β-oxidation of 8:2-FTA, after activation to its CoA-derivative (Dinglasan et al. 2004), the latest studies suggest that 8:2-FTA is transformed to the transient metabolite 7:2-sFTOH, and finally to PFOA by an unknown pathway (Wang et al. 2009). We note, however, that these studies were carried out under very different conditions (degradation

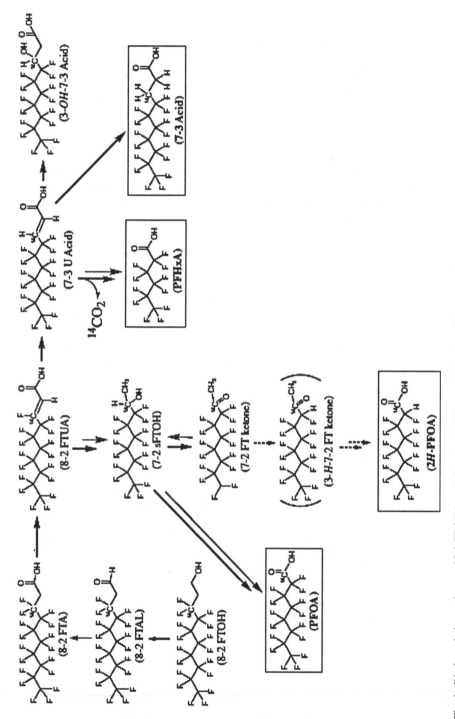

Fig. 1 Biodegradation pathway of 8:2-FTOH in soil. Reprinted with permission from Wang et al. (2009) © Elsevier

defined in mineral medium vs. in soil), so it cannot be excluded that both mechanisms are possible. However, the pathway suggested by Wang et al. (2009) was corroborated by synthesizing a number of transient metabolites and subjecting them to individual degradation tests to delineate the explicit metabolic pathways. The yield of PFOA ranged between 0.5 (Liu et al. 2007) and 25% (Wang et al. 2009), and the amounts formed are certainly dependent on the duration of the degradation tests.

Very recently, these authors discovered yet another stable metabolite derived from 8:2-FTUA degradation, namely, 2H-PFOA (Wang et al. 2009). The detailed metabolic reactions involved herein have not yet been identified.

Fluorotelomer ethoxylates (FTEO) are polyethoxylated fluorotelomer derivatives used mainly as additives in paints and coatings. Their degradation behavior has been briefly analyzed by Schröder (2003), who found that their ethoxylate chain is degraded in length via carboxylic intermediates under aerobic conditions. Since the extent of de-ethoxylation and the formation of potential metabolites were not investigated in these experiments, a detailed study was carried out by us to evaluate the potential for perfluorocarboxylate formation and to identify stable metabolites (Frömel 2010, manuscript in preparation). The results of our study not only confirmed the Schröder results but also showed that biodegradation proceeds by ethoxylate shortening via the respective carboxylates and virtually ceases when the chain length reaches a number of eight intact ethoxylate groups. Furthermore, these compounds were recently detected in wastewater during a study in our laboratory (Frömel 2010, manuscript in preparation). Apparently, further de-ethoxylation is hampered by the proximity of the perfluorinated alkyl chain. The formation of FTOHs and PFCAs was not encountered in these studies.

4 Biodegradation of N-Alkyl Perfluorooctane Sulfonamide Derivatives

N-Alkyl perfluorooctane sulfonamidoethanols (alkyl = methyl, ethyl) are currently used in paper-protecting applications or are reacted with acrylic or methacrylic acid to form the respective polymers (3M 1999; Stock et al. 2004; Begley et al. 2005; Tittlemier et al. 2006). Therefore, they can enter the environment either by direct emissions (e.g., evaporation or leaching of residual unreacted material out of the finished product) or by indirect emissions (e.g., degradation of polymers consisting of N-alkyl perfluorooctane sulfonamidoethanol (FOSE) alcohol building blocks). Whereas the latter, indirect pathway, is currently only notional, direct emissions actually occur, to wit, polymers may contain residual monomers at concentration ranges of up to 1–2%. The global dispersion of N-ethyl perfluorooctane sulfonamide (N-EtFOSA) and N-methyl perfluorooctane sulfonamide (N-MeFOSA) is easily achieved due to gas phase transport (Shoeib et al. 2006). The most exhaustively investigated compound of this class is N-ethyl perfluorooctane sulfonamidoethanol (N-EtFOSE), the biodegradation pathways of which have been determined in

Fig. 2 Proposed biodegradation pathway of *N*-EtFOSE in an aerobic-activated sludge. Reprinted with permission from Rhoads et al. (2008) © American Chemical Society

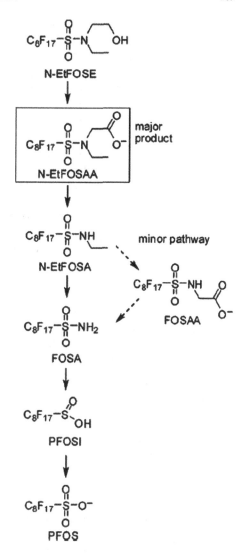

activated sludge (Lange 2000; Rhoads et al. 2008; Sáez et al. 2008). *N*-EtFOSE is rapidly transformed, primarily by the aerobic microorganisms present in activated sludge, as shown in Fig. 2. Similar to FTOHs, the hydroxyl function is initially oxidized to the respective aldehyde and to the carboxylic acid (*N*-ethyl perfluorooctane sulfonamidoacetic acid (N-EtFOSAA)), followed by cleavage of the acetic acid moiety, which leads to *N*-EtFOSA. Other pathways are possible, but are controversial. Although Lange (2000) proposed that *N*-EtFOSA is oxidized to perfluorooctane sulfonamidoacetate (FOSAA) and subsequent degradation to perfluorooctane sulfonamide (FOSA), Rhoads et al. (2008) suggested that this pathway is inferior to direct deamination of *N*-EtFOSA to FOSA. They base their view on individual degradation tests performed on the metabolites and a comparison of transformation rates of these compounds. Also the fate of FOSA is not indisputable: The conversion of FOSA to perfluorooctane sulfinate (PFOSI) has been observed by Lange (2000) and Rhoads et al. (2008); although Lange (2000) observed both PFOS and PFOA formation, Rhoads et al. (2008) only detected PFOS, while PFOA was only present as a contaminant in their PFOSI standard. The amounts of PFOS measured are in the range of 7% at day 35, and PFOA was formed at 0.6%.

In another degradation experiment, N-EtFOSE was exposed to aerobic and anaerobic sludge in a short-term experiment of 96 h (Boulanger et al. 2005). N-EtFOSE was not biotransformed at all under anaerobic conditions, and under aerobic conditions, N-EtFOSAA and PFOSI were the only metabolites detected. N-EtFOSAA was also detected in real wastewater treatment plant (WWTP) influent and effluent samples as well as in the river downstream of this WWTP, which receives domestic and industrial wastewater. The authors suggest that PFOS is not released from wastewater treatment plants by biotransformation of N-EtFOSE.

The anaerobic and aerobic degradation of polyethoxylated N-EtFOSE was investigated and it was concluded that these compounds are degradable only under anaerobic conditions (Schröder 2003). This result is astonishing, because other polyethoxylates, such as nonylphenol ethoxylates or fatty alcohol ethoxylates, are readily degraded consistent with their polyethoxylate chain length (Frömel and Knepper 2008). No metabolites were detected in the samples incubated under anaerobic conditions. These results are unique and have not been confirmed. None of the conditions tested by Schröder achieved a primary biotransformation of the methyl-endcapped N-EtFOSE polyethoxylate, which is consistent with the results obtained for other methyl ethers of polyethoxylates (Schröder 2001). The methyl-endcapping precludes initial oxidation processes of the terminal hydroxyl group, thus degradation cannot take place.

5 The Role of Fluorinated Polymers

One of the main applications of polyfluorinated organic molecules is their use in manufacturing polymers (3M 1999). These polymers comprise perfluorinated polymers such as poly(tetrafluoroethylene) or perfluoropolyethers; fluorotelomer-based polymers such as acrylates, methacrylates, and urethanes (Russell et al. 2008; Washington et al. 2009); and N-EtFOSE or N-methyl perfluorooctane sulfonamidoethanol (N-MeFOSE)-based polymers, which are of the acrylate, urethane, or adipate types (3M 1999).

The number of published articles on the degradation of polymers is still small due to the complexity of the subject and difficulties involved in the experimental setup, since most polymers are water-insoluble. The water-insolubility dramatically affects the timescale of biodegradation, and reproducible results are also difficult to obtain (Eubeler et al. 2009).

Although there has not been any indication that the perfluorinated polymers, such as PTFE, biodegrade, the growing interest and concern about PFOA led to conducting biodegradation studies on fluorotelomer-based polymers. These fluorotelomer-based polymers contain ester functions $R–C(O)O–CH_2–CH_2–R_f$, where R_f is a perfluorinated alkyl chain. Hydrolysis or enzymatic cleavage of the ester would result in free-FTOH, which would then assuredly result in PFCA formation. Because of the high production amounts of these polymers, they may be or may become a significant source of PFOA and other PFCAs in the environment.

Russell et al. (2008) studied the biodegradation of a copolymer being constituted of fluorotelomer acrylate monomers, hydrocarbon acrylate monomers, and dichloroethylene monomers. The resulting polymer had an average molecular weight of 40,000 Da and was incubated in aerobic soil for 2 years. Formation of PFOA and other 8:2-FTOH-related metabolites was observed, but it was shown by mass balance that this is the result of degradation of impurities present in the starting material. The perfluorinated alkyl chain may shield the ester group from esterases or other enzymes that could catalyze the decomposition of the ester into the respective FTOH and the remainder of the polymer (Renner 2008).

However, in another article describing the biodegradation of an acrylate-linked fluorotelomer-based polymer, similar to the one Russell et al. (2008) investigated, it was shown that biodegradation of this polymer did occur releasing several well-known FTOH metabolites (Washington et al. 2009). Thereby, different half-lives of the polymer ranging between 870 and 1400 years were calculated. Assuming that biodegradation of such polymers is surface mediated, the half-life of more finely grained commercial fluorotelomer-based polymers might be as low as 10–17 years. Chemical hydrolysis of this polymer type was not observed within a pH range of 1.2–9 (DuPont Company 2004).

The results reported on polyfluorinated polymers are still controversial (Renner 2008), and new data to confirm the non-degradability are urgently needed to assess the relevance of polymer biodegradation.

Similar to fluorotelomer-based polymers, N-EtFOSE and N-MeFOSE-based polymers may be enzymatically cleaved or hydrolyzed to N-EtFOSE and N-MeFOSE, respectively. The release of these chemicals would then result in PFOS formation. Unfortunately, no studies on biodegradation of such polymers have been yet published.

6 On the Way to Mineralization – Biodegradation of Organic Molecules that Have Low Fluorine Content

In view of the phase out of commercial uses for PFOS and PFOA derivatives, we have recently investigated, in a systematic study, the biodegradability of fluorinated surfactants that have lower fluorine (Peschka et al. 2008a, b). These compounds structurally were ω-substituted alkane-1-sulfonates and had various fluorine-containing endgroups (Table 1), such as a trifluoromethoxy group, a 4-trifluoromethylphenoxy or a bis(trifluoromethyl)amino group. The alkyl chains varied in length between 7 and 13 methylene groups. The objective of this study was to evaluate potentially biodegradable fluorinated functional groups which may be used in the future as novel fluorinated surfactants. We concluded that none of these groups is prone to direct microbial attack. However, biomineralization was achieved for two compounds whose biotransformation processes led to chemically or biologically unstable fluorinated metabolites. The only compound that was completely biomineralized was 10-trifluoromethoxy-decane-1-sulfonate (TFM-DS) (Peschka et al. 2008a). The biotransformation pathway started with desulfonation and oxidation to a carboxylic acid, the carbon chain of which was shortened by successive β-oxidation (Fig. 3). Trifluoromethanol was formed as an intermediate, which is instable in water and thus decays chemically. A second pathway simultaneously led to oxidized derivatives of the parent compound, in which one of the methylene groups was initially oxidized to an alcohol and then to the respective ketone. These carbonyl derivatives decay slowly, but finally are mineralized as well. Therefore, virtually complete biomineralization was achieved for TFM-DS.

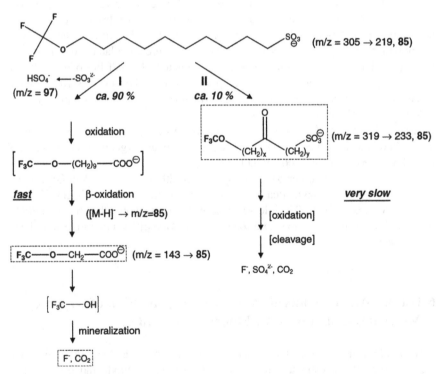

Fig. 3 Biodegradation pathways of TFM-DS. Reprinted with permission from Peschka et al. (2008a) © Elsevier

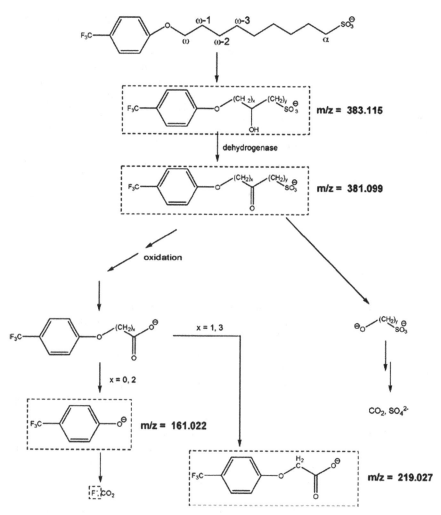

Fig. 4 Proposed biodegradation pathway of TFMP-NS. Structures in frames signify compounds analytically detected. Reprinted with permission from Peschka et al. (2008b) © Elsevier

The compound 9-[4-(trifluoromethyl)phenoxy]-nonane-1-sulfonate (TFMP-NS) showed a somewhat different degradation pathway as illustrated in Fig. 4; degradation was initiated by oxidation to give an alcohol, which was further oxidized to a ketone (Peschka et al. 2008b). Again, several oxidation sites on this structure are available, but the only one confirmed was the ω-oxidation in relation to the sulfonate group. Even in that case mineralization was not complete, since 4-(trifluoromethyl)phenoxy acetic acid was formed during transformation.

For the bis(trifluoromethyl)amino compounds (ω-[bis(trifluoromethyl)amino] alkane-1-sulfonates, BTFMA-AS), which were investigated as a homologous mixture of heptane to tridecane derivatives, it was shown that the alkyl chain length is a crucial factor for desulfonation (Frömel et al. 2008). Complete desulfonation was only achieved when 10 or more methylene groups separate the sulfonate group from the bis(trifluoromethyl)amino group. Furthermore, no mineralization was achieved, because of the emergence of bis(trifluoromethyl)amino acetic acid and 3-[bis(trifluoromethyl)amino] propionic acid, both of which are stable. Apparently, the bis(trifluoromethyl)amino group hinders the β-oxidation before the chemically unstable bis(trifluoromethyl)amine is released. Therefore, no increase in fluoride concentration was detected. This study provided novel fluorinated functional groups that are adequate to substitute for those of the long-chained perfluorinated compounds. The chemical stability of these latter compounds is impressive, which is likely to render them as irreplaceable in some ways. Incorporation of several of the above-mentioned short-chained fluorinated functional groups may provide similar performance to those that have the long-chained perfluorinated alkyl groups.

7 Summary

The incorporation of fluorine into organic molecules entails both positive and adverse effects. Although fluorine imparts positive and unique properties such as water- and oil-repellency and chemical stability, adverse effects often pervade members of this compound class. A striking property of long perfluoroalkyl chains is their very pronounced environmental persistence.

The present review is the first one designed to summarize recent accomplishments in the field of biodegradation of fluorine-containing surfactants, their metabolites, and structural analogs. The pronounced scientific and public interest in these chemicals has given impetus to undertake numerous degradation studies to assess the sources and origins of different fluorinated analog chemicals known to exist in the environment. It was shown that biodegradation plays an important role in understanding how fluorinated substances reach the environment and, once they do, what their fate is.

Today, PFOS and PFOA are ubiquitously detected as environmental contaminants. Their prominence as contaminants is mainly due to their extreme persistence, which is linked to their perfluoroalkyl chain length. It appears that desulfonation of a highly fluorinated surfactants can be achieved if an α-situated H atom, in relation to the sulfonate group, is present, at least under sulfur-limiting conditions.

Molecules that are less heavily fluorinated can show very complex metabolic behavior, as is the case for fluorotelomer alcohols. These compounds are degraded via different but simultaneous pathways, which produce different stable metabolites, one of which is the respective perfluoroalkanoate (8:2-FTOH is transformed to PFOA).

Preliminary screening tests indicate that fluorinated functional groups, such as the trifluoromethoxy group and the p-(trifluoromethyl)phenoxy group, may be useful implementations in novel, environmentally benign fluorosurfactants. More specifically, trifluoromethoxy groups constitute a substitute for those that have been used in the past; this functionality is degradable when it appears in structures that are normally subject to biodegradation. Other compounds tested did not meet this criterion.

Interdisciplinary investigations on fluorinated surfactants are still very much needed and will certainly continue during the next many years. For instance, the role of fluorinated polymers in contributing small fluorinated molecules to the environmental burden still has not been fully understood.

Acknowledgments Tobias Frömel wishes to thank the German Water Chemical Society for the gracious donation of a grant to support his research.

References

3M (1999) Fluorochemical use, distribution and release overview, U.S. Public Docket AR-226-0550. http://www.chemicalindustryarchives.org/dirtysecrets/scotchgard/pdfs/226-0550.pdf. Last access 23 Sep 2009

Begley TH, White K, Honigfort P, Twaroski ML, Neches R, Walker RA (2005) Perfluorochemicals: potential sources of and migration from food packaging. Food Addit Contam 22:1023–1031

Boulanger B, Vargo JD, Schnoor JL, Hornbuckle KC (2005) Evaluation of perfluorooctane surfactants in a wastewater treatment system and in a commercial surface protection product. Environ Sci Technol 39:5524–5530

Butt CM, Mabury SA, Muir DCG, Braune BM (2007) Prevalence of long-chained perfluorinated carboxylates in seabirds from the Canadian arctic between 1975 and 2004. Environ Sci Technol 41:3521–3528

Dinglasan MJA, Ye Y, Edwards EA, Mabury SA (2004) Fluorotelomer alcohol biodegradation yields poly- and perfluorinated acids. Environ Sci Technol 38:2857–2864

DuPont Company (2004) Hydrolytic stability study report. U.S. EPA Administrative Record OPPT2003-0012-2607. www.regulations.gov. Last access 23 Sep 2009

Ellis DA, Martin JW, De Silva AO, Mabury SA, Hurley MD, Andersen MPS, Wallington TJ (2004) Degradation of fluorotelomer alcohols: a likely atmospheric source of perfluorinated carboxylic acids. Environ Sci Technol 38:3316–3321

Ellis DA, Martin JW, Mabury SA, Hurley MD, Andersen MPS, Wallington TJ (2003) Atmospheric lifetime of fluorotelomer alcohols. Environ Sci Technol 37:3816–3820

Eubeler JP, Zok S, Bernhard M, Knepper TP (2009) Environmental biodegradation of synthetic polymers I. Test methodologies and procedures. Trac – Trend Anal Chem 9:1057–1072

Frömel T, Knepper TP (2008) Mass spectrometry as an indispensable tool for studies of biodegradation of surfactants. Trac – Trend Anal Chem 27:1091–1106

Frömel T, Peschka M, Fichtner N, Hierse W, Ignatiev NV, Bauer KH, Knepper TP (2008) ω-(Bis(trifluoromethyl)amino)alkane-1-sulfonates: synthesis and mass spectrometric study of the biotransformation products. Rapid Commun Mass Spectrom 22:3957–3967

Frömel T, Knepper TP (2010) Fluoroteomer ethoxylates: Sources of highly fluorinated environmental contaminants part I: Biotransformation. Chemosphere, in Press

Furdui VI, Stock NL, Ellis DA, Butt CM, Whittle DM, Crozier PW, Reiner EJ, Muir DCG, Mabury SA (2007) Spatial distribution of perfluoroalkyl contaminants in lake trout from the Great Lakes. Environ Sci Technol 41:1554–1559

Giesy JP, Kannan K (2001) Global distribution of perfluorooctane sulfonate in wildlife. Environ Sci Technol 35:1339–1342

Goss KU, Bronner G, Harner T, Hertel M, Schmidt TC (2006) The partition behavior of fluorotelomer alcohols and olefins. Environ Sci Technol 40:3572–3577

Hansen KJ, Clemen LA, Ellefson ME, Johnson HO (2001) Compound-specific, quantitative characterization of organic fluorochemicals in biological matrices. Environ Sci Technol 35:766–770

Key BD, Howell RD, Criddle CS (1997) Fluorinated organics in the biosphere. Environ Sci Technol 31:2445–2454

Key BD, Howell RD, Criddle CS (1998) Defluorination of organofluorine sulfur compounds by *Pseudomonas* sp. strain D2. Environ Sci Technol 32:2283–2287

Lange CC (2000) The aerobic biodegradation of *N*-EtFOSE alcohol by the microbial activity present in municipal wastewater treatment sludge. http://www.nikwax.com/cmsdata/Downloads/pr/5-3 M_biodegradation_report.pdf. Last access 23 Sep 2009

Lau C, Anitole K, Hodes C, Lai D, Pfahles-Hutchens A, Seed J (2007) Perfluoroalkyl acids: a review of monitoring and toxicological findings. Toxicol Sci 99:366–394

Lei YD, Wania F, Mathers D, Mabury SA (2004) Determination of vapor pressures, octanol–air, and water–air partition coefficients for polyfluorinated sulfonamide, sulfonamidoethanols, and telomer alcohols. J Phys Chem A 49:1013–1022

Liao C, Wang T, Cui L, Zhou Q, Duan S, Jiang G (2009) Changes in synaptic transmission, calcium current, and neurite growth by perfluorinated compounds are dependent on the chain length and functional group. Environ Sci Technol 43:2099–2104

Liu J, Lee LS (2005) Solubility and sorption by soils of 8:2 fluorotelomer alcohol in water and cosolvent systems. Environ Sci Technol 39:7535–7540

Liu J, Lee LS, Nies LF, Nakatsu CH, Turcot RF (2007) Biotransformation of 8:2 fluorotelomer alcohol in soil and by soil bacteria isolates. Environ Sci Technol 41:8024–8030

Molina MJ, Rowland FS (1974) Stratospheric sink for chlorofluoromethanes: chlorine atom-catalysed destruction of ozone. Nature 249:810–812

Murphy CD, Schaffrath C, O'Hagan D (2003) Fluorinated natural products: the biosynthesis of fluoroacetate and 4-fluorothreonine in *Streptomyces cattleya*. Chemosphere 52:455–461

Natarajan R, Azerad R, Badet B, Copin E (2005) Microbial cleavage of CF bond. J Fluorine Chem 126:424–435

O'Hagan D, Harper B (1999) Fluorine-containing natural products. J Fluorine Chem 100:127–133

Organisation of Economic Co-operation and Development (1992) OECD guideline for the testing of chemicals 302 B – Zahn-Wellens/EMPA test. http://www.oecd.org/dataoecd/17/18/1948225.pdf. Last access 22 Sep 2009

Organisation of Economic Co-operation and Development (2007) Lists of PFOS, PFAS, PFOA, PFCA, related compounds and chemicals that may degrade to PFCA. http://www.olis.oecd.org/olis/2006doc.nsf/LinkTo/NT00000F9A/$FILE/JT03231059.PDF. Last access 23 Sep 2009

Parsons JR, Sáez M, Dolfing J, de Voogt P (2008) Biodegradation of perfluorinated compounds. Rev Environ Contam Toxicol 196:53–71

Paul AG, Jones KC, Sweetman AJ (2008) A first global production, emission, and environmental inventory for perfluorooctane sulfonate. Environ Sci Technol 43:386–392

Peden-Adams MM, Keller JM, Eudaly JG, Berger J, Gilkeson GS, Keil DE (2008) Suppression of humoral immunity in mice following exposure to perfluorooctane sulfonate. Toxicol Sci 104:144–154

Peschka M, Fichtner N, Hierse W, Kirsch P, Montenegro E, Seidel M, Wilken RD, Knepper TP (2008a) Synthesis and analytical follow-up of the mineralization of a new fluorosurfactant prototype. Chemosphere 72:1534–1540

Peschka M, Fromel T, Fichtner N, Hierse W, Kleineidam M, Montenegro E, Knepper TP (2008b) Mechanistic studies in biodegradation of the new synthesized fluorosurfactant 9-[4-(trifluoromethyl)phenoxy]nonane-1-sulfonate. J Chromatogr A 1187:79–86

Prevedouros K, Cousins IT, Buck RC, Korzeniowski SH (2006) Sources, fate and transport of perfluorocarboxylates. Environ Sci Technol 40:32–44

Remde A, Debus R (1996) Biodegradability of fluorinated surfactants under aerobic and anaerobic conditions. Chemosphere 32:1563–1574

Renner R (2008) Do perfluoropolymers biodegrade into PFOA? Environ Sci Technol 42:648–650

Rhoads KR, Janssen EML, Luthy RG, Criddle CS (2008) Aerobic biotransformation and fate of *N*-ethyl perfluorooctane sulfonamidoethanol (*N*-EtFOSE) in activated sludge. Environ Sci Technol 42:2873–2878

Rowland FS (2002) Stratospheric ozone in the 21st century: the chlorofluorocarbon problem. Environ Sci Technol 25:622–628

Russell MH, Berti WR, Szostek B, Buck RC (2008) Investigation of the biodegradation potential of a fluoroacrylate polymer product in aerobic soils. Environ Sci Technol 42:800–807

Sáez M, de Voogt P, Parsons JR (2008) Persistence of perfluoroalkylated substances in closed bottle tests with municipal sewage sludge. Environ Sci Pollut Res 15:472–477

Schröder HF (2001) Tracing of surfactants in the biological wastewater treatment process and the identification of their metabolites by flow injection-mass spectrometry and liquid chromatography–mass spectrometry and -tandem mass spectrometry. J Chromatogr A 926:127–150

Schröder HF (2003) Determination of fluorinated surfactants and their metabolites in sewage sludge samples by liquid chromatography with mass spectrometry and tandem mass spectrometry after pressurised liquid extraction and separation on fluorine-modified reversed-phase sorbents. J Chromatogr A 1020:131–151

Schultz MM, Barofsky DF, Field JA (2004) Quantitative determination of fluorotelomer sulfonates in groundwater by LC MS/MS. Environ Sci Technol 38:1828–1835

Shoeib M, Harner T, Vlahos P (2006) Perfluorinated chemicals in the arctic atmosphere. Environ Sci Technol 40:7577–7583

Stock NL, Lau FK, Ellis DA, Martin JW, Muir DCG, Mabury SA (2004) Polyfluorinated telomer alcohols and sulfonamides in the North American troposphere. Environ Sci Technol 38: 991–996

Taves DR (1968) Evidence that there are two forms of fluoride in human serum. Nature 217: 1050–1051

Tittlemier SA, Pepper K, Edwards L (2006) Concentrations of perfluorooctanesulfonamides in Canadian total diet study composite food samples collected between 1992 and 2004. J Agric Food Chem 54:8385–8389

United States Environmental Protection Agency (2008) Fate, transport and transformation test guidelines – OPPTS 835.3215 – Inherent biodegradability – Conclawe test. http://www.epa.gov/opptsfrs/publications/835_3215.pdf. Last access 23 Sep 2009

Wallington TJ, Hurley MD, Xia J, Wuebbles DJ, Sillman S, Ito A, Penner JE, Ellis DA, Martin J, Mabury SA, Nielsen OJ, Andersen MPS (2006) Formation of $C_7F_{15}COOH$ (PFOA) and other perfluorocarboxylic acids during the atmospheric oxidation of 8:2 fluorotelomer alcohol. Environ Sci Technol 40:924–930

Wang N, Szostek B, Buck RC, Folsom PW, Sulecki LM, Capka V, Berti WR, Gannon JT (2005a) Fluorotelomer alcohol biodegradation – direct evidence that perfluorinated carbon chains breakdown. Environ Sci Technol 39:7516–7528

Wang N, Szostek B, Buck RC, Folsom PW, Sulecki LM, Gannon JT (2009) 8-2 Fluorotelomer alcohol aerobic soil biodegradation: pathways, metabolites, and metabolite yields. Chemosphere 75:1089–1096

Wang N, Szostek B, Folsom PW, Sulecki LM, Capka V, Buck RC, Berti WR, Gannon JT (2005b) Aerobic biotransformation of 14C-labeled 8-2 telomer B alcohol by activated sludge from a domestic sewage treatment plant. Environ Sci Technol 39:531–538

Washington JW, Ellington JJ, Jenkins TM, Evans JJ, Yoo H, Hafner SC (2009) Degradability of an acrylate-linked, fluorotelomer polymer in soil. Environ Sci Technol 43:6617–6623

Perfluorinated Substances in Human Food and Other Sources of Human Exposure

Wendy D'Hollander, Pim de Voogt, Wim De Coen, and Lieven Bervoets

Contents

1 Introduction

Perfluorinated compounds (PFCs) are ubiquitous environmental contaminants, which persist and may bioaccumulate through the food chain (Haukås et al. 2007; Martin et al. 2004b; Taniyasu et al. 2003). As a consequence, several PFCs have been detected in different biota worldwide. In recent years, an increasing number of papers report high levels of PFCs in blood, tissues, and breast milk from both occupationally and non-occupationally exposed human populations (Kannan et al. 2004; Kärrman et al. 2007; Olsen et al. 2007). The most important exposure pathways of perfluorinated compounds for humans are thought to be intake of drinking water and food and inhalation of dust (Björklund et al. 2009; Ericson et al. 2008a).

W. D'Hollander (✉)

Laboratory for Ecophysiology, Biochemistry and Toxicology, Department of Biology, University of Antwerp, 2020 Antwerp, Belgium

e-mail: wendy.dhollander@ua.ac.be

P. de Voogt (ed.), *Reviews of Environmental Contamination and Toxicology*,
Reviews of Environmental Contamination and Toxicology 208,
DOI 10.1007/978-1-4419-6880-7_4, © Springer Science+Business Media, LLC 2010

Due to the widespread distribution, environmental degradation, and metabolism of the PFCs released into the environment, a very complex exposure situation exists (Fromme et al. 2007a). As a result, the relative contribution to human exposure from different routes or from a single source (e.g., diet) is not yet known. More specifically, it is currently unknown as to what extent exposure to drinking water, food, or dust contributes to the PFCs measured in human breast milk and blood. Moreover, data on levels of PFCs in the human diet are rather scarce (Kärrman et al. 2009; Tittlemier et al. 2006, 2007). Only PFC levels in fish appear to be well documented (Houde et al. 2006). Few studies, however, report the levels of PFCs in drinking water or human food such as vegetables, meat, and eggs (FSA 2006; United States Environmental Protection Agency (US EPA) 2001). Food processing such as cooking (boiling, baking, or grilling) could alter the concentration of PFCs in food and as a consequence affect the risk for humans.

In this review, we summarize the existing data on the levels of perfluorinated compounds that appear in the human diet and in drinking water. Additionally, an overview will be given on what data are available in the literature on PFCs in indoor (house) dust. These overviews are designed to provide a current picture of the contributions PFCs make to human residue burdens by major routes of exposure.

Data were mainly obtained from peer-reviewed articles published up to March 2009. The literature search was performed through use of the web databases ISI Web of Knowledge and Science Direct. Furthermore, data from reports (FSA 2006; US EPA 2001; Schrap et al. 2004; Van den Heuvel-Greve et al. 2006) are included in the review. The collected data consisted mainly of PFC levels expressed as mean concentrations or data ranges.

This chapter will be divided into four parts. In the first part, concentrations of PFCs in edible aquatic species are listed. In the second part the data on PFCs in composite food samples, vegetables, meat, and other food are reviewed. Part three gives the data on levels in drinking water, and finally, in part four the levels in indoor dust are reported.

2 PFCs in Edible Fish and Seafood

It has been well documented that perfluorooctanoic acid (PFOA) and perfluorooctane sulfonate (PFOS) may bioaccumulate in fish and that accumulation of PFCs with chain lengths of C_8–C_{15} in fish tends to increase with increasing chain length (Martin et al. 2003b, 2004b; de Vos et al. 2008). Therefore, fish are potentially an important dietary source of PFCs for humans. In general, muscle is the main part of the fish body that is consumed by humans. Considering the fact that humans prefer muscle tissue, we mainly focus in this chapter on levels of perfluorinated compounds in such fish tissue. A summary of perfluorinated compounds, especially PFOS, PFOA, perfluorononanoic acid (PFNA) and perfluorooctanesulfonamide (PFOSA), that have been found in fish muscle and/or in whole fish homogenates of edible species is illustrated in Table 1. Among PFC fish contaminants, PFOS is the most

Table 1 Mean concentrations (±SD) (ng/g wet weight; wwt) of perfluorinated substances (PFCs) in edible fish species

Species (n) (common name)	Sample location, year(s) of collection		PFOS*	PFOA	PFNA	PFOSA	References
Whole fish homogenates							
Salvelinus namaycush (56) (lake trout)	Lake Ontario, North America, Canada, 1980–2001	Mean	170 ± 43	1.0 ± 0.1	4.2 ± 3.0	16 ± 9.6	Martin et al. (2004b)
Salvelinus namaycush (10) (lake trout)	Lake Ontario, North America, Canada, 2001	Mean	46 ± 5	1.5 ± 0.4	1.1 ± 0.2	0.8 ± 0.1	Furdui et al. (2007)
Salvelinus namaycush (10) (lake trout)	Lake Superior, North America, 2001	Mean	4.8± 0.4	1.1 ± 0.2	1.0 ± 0.2	0.3 ± 0.0	Furdui et al. (2007)
Salvelinus namaycush (10) (lake trout)	Lake Michigan, North America, 2001	Mean	16 ± 3	4.4 ± 1.6	0.6 ± 0.2	1.0 ± 0.3	Furdui et al. (2007)
Salvelinus namaycush (10) (lake trout)	Lake Huron, North America, Canada, 2001	Mean	39 ± 10	1.6 ± 0.3	2.8 ± 0.5	1.6 ± 0.3	Furdui et al. (2007)
Salvelinus namaycush (10) (lake trout)	Lake Erie, North America, Canada, 2001	Mean	121 ± 14	1.6 ± 0.7	2.9 ± 0.4	2.1 ± 0.5	Furdui et al. (2007)
Osmerus mordax (30) (rainbow smelt)	Lake Ontario, North America, Canada, 2002	Mean	110 ± 55	2.0 ± 1.1	6.8 ± 5.1	72 ± 23	Martin et al. (2004b)

Table 1 (continued)

Species (n) (common name)	Sample location, year(s) of collection		PFOS*	PFOA	PFNA	PFOSA	References
Alosa pseudoharengus (12) (alewife)	Lake Ontario, North America, Canada 2002	Mean	46 ± 15	1.6 ± 1.5	0.8 ± 0.8	4.0 ± 3.2	Martin et al. (2004b)
Solea solea (p; 1 kg) (Sole)	Western Scheldt, Netherlands, 2005	Mean	120	/	/	/	Van den Heuvel-Greve et al. (2006)
Ammodytes sp. (p; 1 kg) (sandeel)	Western Scheldt, Netherlands, 2005	Mean	50	/	/	/	Van den Heuve-Greve et al. (2006)
Boreogadus saida (6) (Arctic cod)	Davis strait, Artic Ocean, 2000–2001	Mean	1.3 ± 0.7	0.16 ± 0.06	/	nd	Tomy et al. (2004b)
Mixture of whole fish[a]	Missouri River, USA, 2005	Mean	84.7	<1	0.43	/	Ye et al. (2008)
Mixture of whole fish[a]	Mississippi River, USA, 2005	Mean	83.1	<0.2	0.78	/	Ye et al. (2008)
Mixture of whole fish[a]	Ohio River, USA, 2005	Mean	147	<1	1.0	/	Ye et al. (2008)
Muscle							
Trisopterus luscus (4) (bib)	North Sea Belgium 2001	Range	<10–30	/	/	/	Hoff et al. (2003)
Trisopterus luscus (17) (bib)	Western Scheldt, Netherlands, 2001	Range	<10–107	/	/	/	Hoff et al. (2003)
Pleuronectes platessa (18) (plaice)	North Sea Belgium 2001	Range	<10–39	/	/	/	Hoff et al. (2003)

Table 1 (continued)

Species (n) (common name)	Sample location, year(s) of collection		PFOS*	PFOA	PFNA	PFOSA	References
Pleuronectes platessa (4) (plaice)	Western Scheldt, Netherlands, 2001	Range	<10–87	/	/	/	Hoff et al. (2003)
Anguilla anguilla (p: 5 kg) (European eel)	Western Scheldt, Netherlands, 2005	Range	62–110	/	/	/	Van den Heuvel-Greve et al. (2006)
Anguilla anguilla (300) (European eel)	Rivers, Netherlands, 2005	Range	<8–143	/	/	/	Sharp et al. (2004)
Sprattus sprattus (p: 3.5 kg) (sprat)	Western Scheldt, Netherlands, 2005	Mean	25	/	/	/	Van den Heuvel-Greve et al. (2006)
Platichthys flesus (European flounder)	Western Scheldt, Netherlands, 2005	Mean	23	/	/	/	Scharp et al. (2004)
Platichthys flesus (European flounder)	Wadden Sea, Netherlands, 2005	Mean	<3.7	/	/	/	Scharp et al. (2004)
Micropterus dolomieu (smallmouth bass)	Rivers of Michigan, USA, 1998–1999	Mean	9	<2	/	1.6	Kannan et al. (2005)
Carp	Rivers of Michigan, USA, 1998–1999	Mean	124	<36	/	<19	Kannan et al. (2005)
Carp (10)	Saginaw Bay, Michigan, USA, 1990	Mean	120	/	/	/	Giesy and Kannan (2001)
Chinook salmon (6)	Michigan waters, USA, 1990	Mean	110	/	/	/	Giesy and Kannan (2001)

Table 1 (continued)

Species (n) (common name)	Sample location, year(s) of collection		PFOS*	PFOA	PFNA	PFOSA	References
Lake whitefish (5)	Michigan waters, USA, 1990	Mean	130	/	/	/	Giesy and Kannan (2001)
Brown trout (10)	Michigan waters, USA, 1990	Mean	/	/	/	/	Giesy and Kannan (2001)
Pelagic fish[b]	Mediterranean Sea unknown	Mean	4	/	/	/	Nania et al. (2009)
Benthonic fish[c]	Mediterranean Sea unknown	Mean	13	/	/	/	Nania et al. (2009)
Pseudosciaena plyacti (2) (small yellow croaker)	Guangzhou, China, 2004	Mean	2.93	<0.25	<0.25	/	Gulkowska et al. (2006)
Pseudosciaena plyacti (2) (small yellow croaker)	Zhoushan, China, 2004	Mean	0.92	<0.25	<0.25	/	Gulkowska et al. (2006)
Pampus argenteus (2) (silvery pomfret)	Guangzhou, China, 2004	Mean	0.67	<0.25	<0.25	/	Gulkowska et al. (2006)
Pampus argenteus (2) (silvery pomfret)	Zhoushan, China, 2004	Mean	0.38	<0.25	<0.25	/	Gulkowska et al. (2006)
Trichiurus haumela (2) (belt fish)	Guangzhou, China, 2004	Mean	0.91	<0.25	<0.25	/	Gulkowska et al. (2006)

Table 1 (continued)

Species (n) (common name)	Sample location, year(s) of collection		PFOS*	PFOA	PFNA	PFOSA	References
Pneumatophorus japonicus (2) (Japanese mackerel)	Guangzhou, China, 2004	Mean	2.18	<0.25	<0.25	/	Gulkowska et al. (2006)
Argyrosomus argentatu (2) (whitemouth croaker)	Zhoushan, China, 2004	Mean	0.86	<0.25	<0.25	/	Gulkowska et al. (2006)
Muraenesox cinereus (2) (conger pike)	Zhoushan, China, 2004	Mean	1.77	<0.25	<0.25	/	Gulkowska et al. (2006)

*PFOS: perfluorooctane sulfonate; PFOA: perfluorooctanoic acid; PFNA: perfluorononanoic acid; PFOSA: perfluorooctanesulfonamide

p: pooled sample; nd: not detected; /: not analyzed

[a]Mixture of small and large fish. Small fish sample was a composite of 20–200 individuals of one species whose adults are small. Large fish sample included 3–5 individuals. For more details on the species composition see Ye et al. (2008)

[b]*Mugil cephalus* (gray mullet), *Dentex dentex* (common dentex), *Trachurus mediterraneus* (horse mackerel), *Lamna nasus* (porbeagle), *Mustelus mustelus* (common smooth-hound), *Xiphias gladius* (sword fish), *Thunnus thynnus* (tuna)

[c]*Conger conger* (European conger), *Scyliorhinus canicula* (small-spotted catshark), *Mullus surmuletus* (red mullet), *Pagellus erythrinus* (common Pandora), and *Scorpaena scrofa* (large-scaled scorpion fish)

crucial and prominent compound found. Reports suggest no considerable differences in PFC concentrations among freshwater and marine fish species. However, the highest mean PFOS concentration (170 ng/g wet weight (wwt)) was detected in lake trout, collected from Lake Ontario (Martin et al. 2004b). Other PFOS concentrations in lake trout from the Great Lakes ranged from 16 (Lake Michigan) to 121 ng/g wwt (Lake Erie) (Furdui et al. 2007).

The PFOS concentration in lake trout from Lake Ontario increased significantly from 43 to 180 ng/g wwt in the period 1980–2001 (Martin et al. 2004b). This temporal trend was not confirmed by the study of Furdui et al. (2007), wherein an average PFOS concentration of 46 ng/g wwt was measured in lake trout collected from Lake Ontario, in 2001.

Hitherto, only limited data were available from the same species collected at the same site in different periods; therefore, it is very difficult to draw conclusions on temporal changes of PFC contaminants in fish tissue.

PFOA is the second most perfluorinated compound that is most often subjected to analysis in fish. It has been shown that PFOA is detected at much lower concentrations than is PFOS. As a consequence, in most studies the concentrations of PFOA recorded in fish tissue remained below the detection limit. However, quantifiable concentrations were detected in lake trout (Martin et al. 2004b; Furdui et al. 2007), rainbow smelt (Martin et al. 2004b), and alewife, with concentrations ranging from 0.16 to 6.8 ng/g wwt. The difference between the observed PFOS and PFOA concentrations in fish suggests a lower potential of PFOA to bioaccumulate in fish as compared to PFOS. This observation was further confirmed by laboratory experiments, which revealed a 1,000-fold lower bioconcentration factor for PFOA compared to PFOS (Martin et al. 2003b; Gruber et al. 2007).

A restricted number of studies also reported perfluorohexane sulfonate (PFHxS), perfluoroundecanoic acid (PFUnA), perfluorodecanoic acid (PFDA), perfluoroheptanoic acid (PFHpA), and perfluorohexanoic acid (PFHxA) in fish muscle (Giesy and Kannan 2001; Kannan et al. 2005; Gulkowska et al. 2006; Nakata et al. 2006; Furdui et al. 2007; Ye et al. 2008), but these compounds were found above the detection limit in only a few cases. For example, Ye et al. (2008) detected PFHxS at a maximum concentration of 1.89 ng/g wwt in a mixture of whole fish in the Missouri River, USA. Concentrations of the other PFCs analyzed in this study were all generally at least one order of magnitude lower than PFOS levels and had an overall median concentration of 3.71 (PFHxA), 0.82 (PFDA), and 0.36 ng/g (PFHxS) wwt. Notably, high concentrations of the other PFCs in fish were also broadly dispersed among the rivers investigated, including PFHxS at 8.14 ng/g (Missouri), PFUnA at 48.0 ng/g (Mississippi), and PFDA at 9.01 ng/g and PFHxA at 18.4 ng/g wwt, both from the Ohio River (Ye et al. 2008). In whole body homogenates of lake trout collected from the Great Lakes in 2001, the highest values found were 4.9 ng/g wwt for PFDA, 3.5 ng/g wwt each for PFUnA and perfluorotridecanoic acid (PFTrA), and 9.8 ng/g wwt for perfluorodecane sulfonate (PFDS). An overview of analyses from long-chain perfluorinated acids (= C >10) in fish is given in Table 2.

Martin et al. (2004b) detected relatively high mean concentrations of the longer-chain PFCs in fish collected from Lake Ontario, Canada. The highest concentration

Table 2 Mean concentrations (ng/g wwt) of long-chain perfluorinated substances in edible fish species

Species (n) (common name)	Sample location, year collection	PFDA*	PFUnA	PFDoA	PFTrA	PFTeA	PFPA	References
Salvelinus namaycush (56) (lake trout)	Lake Ontario, Canada,1980–2001	6.1	8.3	3.9	4.6	1.3	<0.5	Martin et al. (2004b)
Salvelinus namaycush (50) (lake trout)	Great Lakes[a], Canada, 1980–2001	2.1	1.9	0.7	2.2	1.1	0.6	Furdui et al. (2007)
Mixture of whole fish[c] (20)	Rivers[b], USA, 2005	1.79	3.5	0.8	/	/	/	Ye et al. (2008)
Osmerus mordax (30) (rainbow smelt)	Lake Ontario, Canada, 2002	6.1	7.0	3.9	3.8	0.6	< 0.5	Martin et al. (2004b)
Alosa pseudoharengus (12) (alewife)	Lake Ontario, Canada, 2002	1.4	1.3	2.1	1.5	< 0.5	< 0.5	Martin et al. (2004b)

l: not analyzed; < x: x = detection limit

*PFDA: perfluorodecanoic acid; PFUnA: perfluoroundecanoic acid; PFDoA: perfluorododecanoic acid; PFTrA: perfluorotridecanoic acid; PFTeA: perfluorotetradecanoic acid; PFPA: perfluoropentadecanoic acid

[a]Mean of all investigated lakes: Lake Superior, Lake Michigan, Lake Huron, Lake Erie, and Lake Ontario

[b]Mean values of all investigated rivers: Missouri River, Mississippi River, Ohio River

[c]Mixture of small and large fish. Small fish sample was a composite of 20–200 individuals of one species whose adults are small. Large fish sample included 3–5 individuals. For more details on the species composition see Ye et al. (2008)

of these PFCs was 8.3 ng/g wwt for PFUnA. These authors concluded that individual PFCs were generally detected at lower concentrations than were PFOS, and total PFOS equivalents (PFOS and PFOSA) exceeded the sum of all PFCs by a factor of between 1.8 and 12 within each species analyzed. This pattern was similar to the relative abundance of both fluorochemical classes in Arctic animals (Martin et al. 2004a) and in lake trout collected from the Great Lakes (Furdui et al. 2007).

Tomy et al. (2004b) detected a relatively high mean concentration (92.8 ng/g wwt) of N-ethyl perfluorooctane sulfonamide (N-EtPFOSA) in Arctic cod, ranging between 9.6 and 144.6 ng/g wwt. Since transformation of N-EtPFOSA to PFOS and PFOSA by rainbow trout microsomes has been reported (Tomy et al. 2004a), N-EtPFOSA is an important compound to measure in biota and in human samples. However, up to now, the available data on N-EtPFOSA levels in living biota have been limited.

Although several authorities recommend not eating fish liver because of the risk associated with high intake of persistent organic pollutants (POPs), fish liver (and oil) is still consumed in a limited number of countries (e.g., Norway, Japan, and Iceland). Therefore, it is interesting to note that PFC levels in liver are at least two orders of magnitude higher than exists in muscle tissue (Nania et al. 2009). In liver of tuna from the Mediterranean Sea, PFOS concentrations up to 87 ng/g wwt were detected (Kannan et al. 2002). In Japan, concentrations of total PFCs in skipjack tuna livers ranged from <1 to 83 ng/g wwt (Hart et al. 2008). PFOS and PFUnA were the prominent compounds found.

Similar to fish, PFOS is the dominant PFC found in edible aquatic invertebrates such as shrimp, mussels, clams, and oysters (Table 3). Extremely high values were measured in shrimp collected in 2001 from the North Sea and the Western Scheldt (Belgium), which had mean PFOS concentrations of 139 and 215 ng/g wwt, respectively (Van de Vijver et al. 2003). However, in 2005, in samples collected from the same sites, the mean PFOS concentration had decreased to 29 ng/g wwt in shrimp from the Western Scheldt and was below the detection limit in shrimp from the North Sea (Schrap et al. 2004; Van den Heuvel-Greve et al. 2006). It is worth noting that, in the latter study, peeled rather than unpeeled shrimp were analyzed; however, it seems unlikely that this is the reason for the observed difference. Levels of PFOS in shrimp from the Western Scheldt were much higher compared to levels in shrimp collected from the Arctic Ocean and in China (i.e., 0.35 and 7.42 ng/g wwt, respectively (Tomy et al. 2004b; Gulkowska et al. 2006)).

A few papers report on PFC levels in bivalves. Kannan et al. (2002) measured PFOS concentrations up to 1,225 ng/g dry weight (dwt) in oysters collected in the Gulf of Mexico and in the Chesapeake Bay in the USA. Other studies on oysters reported much lower concentrations (6.0 and <1 ng/g wwt) from the Ariake Sea (Nakata et al. 2006) and China (Gulkowska et al. 2006), respectively. Cunha et al. (2005) measured PFOS in mussels from several estuaries in the North of Portugal. They detected high concentrations in all the investigated estuaries; the minimum level found was 36.8 and the maximum was 126.0 ng/g wwt. With the exception of the PFOS concentration measured in Portugal (Cunha et al. 2005), the average concentration (0.54 ng/g wwt) in mussels from this location was similar to that found

Table 3 Mean, minimum, and maximum concentrations (ng/g wwt) of PFOS, PFOA, PFNA, and PFOSA in edible aquatic invertebrates

Species (n) (common name)	Sample location, year collected		PFOS	PFOA	PFNA	PFOSA	References
Crangon crangon (17) (shrimp)	Western Scheldt, Belgium, 2001	Mean	215	/	/	/	Van de Vijver et al. (2003)
		Min	66				
		Max	518				
Crangon crangon (29) (shrimp)	North Sea Belgium, 2001	Mean	139	/	/	/	Van de Vijver et al. (2003)
		Min	19				
		Max	519				
Crangon crangon (p; 4 kg) ("peeled" shrimp)	Western Scheldt, Belgium, 2005	Mean	29	/	/	/	Van den Heuvel-Greve et al. (2006)
		Min	/				
		Max	/				
Crangon crangon (p; 1 kg) (shrimp)	North Sea, Netherlands, 2005	Mean	< 5	< 2.5	/	/	Schrap et al. (2004)
		Min	/				
		Max	/				
Pandalus borealis and Hymenodora glacialis (7) (shrimp)	Davis Strait, Artic Ocean, 2000–2001	Mean	0.35	0.17	/	nd	Tomy et al. (2004b)
		Min	nd	nd			
		Max	0.9				
Oratosquilla oratoria (2) and Metapenaeus ensis (2) (shrimp)	Guangzhou, China, 2004	Mean	7.42	0.44	<0.25	/	Gulkowska et al. (2006)
		Min	/	/			
		Max	/	/			
Oratosquilla oratoria (2) and Metapenaeus ensis (2) (shrimp)	Zhoushan, China, 2004	Mean	1.54	/	<0.25	/	Gulkowska et al. (2006)
		Min	/				
		Max	/				
Panulirus stimpsoni (2) (Chinese spiny lobster)	Guangzhou, China, 2004	Mean	1.83	0.31	<0.25	/	Gulkowska et al. (2006)
		Min	/	/	/		
		Max	/	/			

Table 3 (continued)

Species (n) (common name)	Sample location, year collected		PFOS	PFOA	PFNA	PFOSA	References
Shellfish							
Crassostrea virginica (1540) (oyster)	Gulf of Mexico and Chesapeake Bay, USA, 1996–1998	Mean	387[a]	/	/	/	Kannan et al. (2002)
		Min	<45[a]				
		Max	1225[a]				
Oyster (5)	Ariake Sea, Japan, 2003	Mean	<0.3	6.0	<1.5	<1.5	Nakata et al. (2006)
		Min	/	3.4	/	/	
		Max	/	81	/	/	
Ostrea (2) (oysters)	Guangzhou, China, 2004	Mean	0.54	0.31	<0.25	/	Gulkowska et al. (2006)
		Min	/	/	/	/	
		Max	/	/	/	/	
Perna viridis (2) (mussel)	Guangzhou, China, 2004	Mean	0.42	0.31	<0.25	/	Gulkowska et al. (2006)
		Min	/	/	/	/	
		Max	/	/	/	/	
Perna viridis (2) (mussel)	Zhoushan, China, 2004	Mean	0.42	0.48	<0.25	/	Gulkowska et al. (2006)
		Min	/	/	/	/	
		Max	/	/	/	/	
Mussel (5)	Ariake Sea, Japan, 2003	Mean	<0.3	9.5	1.6	<1.5	Nakata et al. (2006)
		Min	/	6.3	<1.5	/	
		Max	/	14	2.1	/	
Mussel (p;1 kg)	Wadden Sea & Eastern Scheldt, Netherlands, 2005	Mean	<3	<2.6	/		Schrap et al. (2004)
		Min	/	/			
		Max	/	/			
Mytilus edulis (400) (blue mussel)	Mediterranean Sea, unknown	Mean	<2	<1.5	/	/	Nania et al. (2009)
		Min	<2	<1.5			
		Max	3	2.5			
Mytilus galloprovincialis (100) (mussel)	10 North-central estuaries Portugal, not given	Mean	72.5	/	/	/	Cunha et al. (2005)
		Min	36.8				
		Max	125.9				

Table 3 (continued)

Species (n) (common name)	Sample location, year collected		PFOS	PFOA	PFNA	PFOSA	References
Dreissena polymorpha (>100) (zebra mussel)	Rivers of Michigan, USA, 1998–1999	Mean	/	<5	/	/	Kannan et al. (2005)
		Min	< 2			<2	
		Max	3.1			2.7	
Gomphinia aequilatera (2) and Meretrix lusoria (2) venus	Guangzhou, China, 2004	Mean	0.38	0.27	<0.25	/	Gulkowska et al. (2006)
		Min	/	/			
		Max	/	/			
Gomphinia aequilatera (2) venus	Zhoushan, China, 2004	Mean	0.51	0.29	<0.25	/	Gulkowska et al. (2006)
		Min	/	/			
		Max					
Cerastoderma edule (p; 1 kg) (cockle)	North Sea, Belgium, 2005	Mean	ND	/	/	/	Van den Heuvel-Greve et al. (2006)
		Min	/				
		Max					
Clam (6)	Ariake Sea, Japan, 2003	Mean	<0,3	7,5	< 1,5	< 1,5	Nakata et al. (2006)
		Min	/	<0,3			
		Max	/	11			
Mya truncata and Serripes groenlandica (5) (clam)	Davis Strait, Artic Ocean, 2000–2001	Mean	0.28	nd	/	nd	Tomy et al. (2004b)
		Min	0.08		/		
		Max	0.6		/		
Venus sp. (400) (clam)	Mediterranean Sea, not given	Mean	< 2	15	/	/	Nania et al. (2009)
		Min	< 2	12	/	/	
		Max	3	16	/	/	
Loligo vulgaris (30) (squid)	Mediterranean Sea, not given	Mean	3	2.5	/	/	Nania et al. (2009)
		Min	< 2	<1.5	/	/	
		Max	3	40	/	/	
Ommastrephidae (2) (squid)	Guangzhou, China, 2004	Mean	1.07	0.43	/	/	Gulkowska et al. (2006)
		Min	/	/	/	/	
		Max	/	/	/	/	

Table 3 (continued)

Species (*n*) (common name)	Sample location, year collected		PFOS	PFOA	PFNA	PFOSA	References
Ommastrephidae (2) (squid)	Zhoushan, China, 2004	Mean	1.32	0.35	/	/	Gulkowska et al. (2006)
		Min	/	/	/	/	
		Max	/	/	/	/	
Sepiidae (2) (cuttlefish)	Guangzhou, China, 2004	Mean	0.87	0.31	<0.25	/	Gulkowska et al. (2006)
		Min	/	/			
		Max	/	/			
Sepiidae (2) (cuttlefish)	Zhoushan, China, 2004	Mean	0.96	0.31	<0.25	/	Gulkowska et al. (2006)
		Min	/	/			
		Max	/	/			
Crayfish	Rivers of Michigan, USA, 1998–1999	Mean	/	<0.2	/	/	Kannan et al. (2005)
		Min	2.4	/	<1		
		Max	4.3	/	1.6		

[a]Dry weight; p: pooled sample; /: not analyzed; < x: x = detection limit; nd: not detected

in oysters collected in China and Japan. The PFOA values in mussels were generally in the same range as was PFOS, except for the mussels from the Ariake Sea in Japan, in which the PFOA concentration was 10 times higher (9.5 ng/g wwt; Nakata et al. 2006). In the Nania et al. (2009) study, higher PFOA than PFOS levels were found in clam but comparable levels were found in mussels, which was attributed to differences in habitat and feeding behavior. In contrast, Gulkowska et al. (2006) reported lower PFOA than PFOS concentrations in clams (*Venus* sp.).

Generally, the concentrations in the other molluscs follow the same trend as observed for mussels and oysters, i.e., PFC concentrations are lower than 1 ng/g wwt. This is consistent with the observations made in China, Belgium, Japan, and in the Artic Ocean. Only in *Venus* sp., collected from the Mediterranean Sea, was the mean PFOA concentration higher (15 ng/g wwt). In squid and cuttlefish, the measured PFOS and PFOA concentrations were slightly higher than those observed in bivalves; the observed range was between 0.31 and 3 ng/g wwt (Gulkowska et al. 2006; Nania et al. 2009).

Unfortunately, data on other PFCs in edible invertebrates are rare. Gulkowska et al. (2006) did find the following PFC components: PFHxS, PFUnA, PFDA, PFHpA, and PFHxA. Among perfluorocarboxylic acids, residues of PFOA and PFUnA were the most prominent compounds found in seafood. Quantifiable concentrations of PFOA were found in all crab and mollusc tissues and two shrimp species at a mean concentration of 0.48 ng/g wwt. PFUnA was found at a slightly greater mean concentration of 0.52 ng/g wwt (Gulkowska et al. 2006). This trend was in accord with the bioaccumulation potential (BCP) of perfluoroalkyl carboxylates (PFCAs), whereby bioaccumulation increases with increasing perfluoroalkyl chain length (Martin et al. 2003a, b). The widespread occurrence of long-chain PFCAs has been examined in the Canadian Artic (Martin et al. 2004a). The contamination profile for fish was similar to that observed by Gulkowska et al. (2006); the dominant PFCA in all fish was PFUnA, but lower concentrations were recorded for both longer- and shorter-chain homologs. Tomy et al. (2004b) detected relatively high mean concentrations of *N*-EtPFOSA in clams and in shrimp of, respectively, 20.1 and 10.4 ng/g wwt.

In general, concentrations of PFCs are expected to increase with increasing trophic level. This trend has been observed in the Great Lakes food chain (Kannan et al. 2005). However, higher concentrations of perfluoroalkyl contaminants were reported in lower trophic levels in seafood from China (Gulkowska et al. 2006) and in invertebrate species from Lake Ontario (Martin et al. 2004b). Several literature findings suggest that benthic organisms do not biomagnify PFCs to the extent that pelagic ones do, but so far, the overall picture is still inconclusive. Martin et al. (2004b) observed strong associations between PFC levels in pelagic species (*Mysis*, alewife, smelt, and lake trout) and trophic level, if benthic species (*Diporeia* and sculpin) were excluded from the regression analysis. Kidd et al. (2001) previously demonstrated that there is a divergence between pelagic and benthic-source biomagnification for persistent organochlorines due to the higher carbon turnover rate of benthic species. Martin et al. (2004b) observed statistically significant trends in the pelagic food web, whereby PFOS, PFDA, PFUnA, and PFTrA concentrations increased with increasing trophic level, suggesting that biomagnification occurred. In the benthic macroinvertebrate *Diporeia*, however, which occupied the lowest trophic level of all the organisms analyzed, the highest mean concentration of each perfluorinated contaminant was detected. This suggests that the major source of PFCs was the sediment, not the water (Martin et al. 2004b). This may be a result of the sorption of perfluorinated acids or neutral perfluoroalkyl substances to organic matter, followed by sedimentation and subsequent uptake by benthic invertebrates. Sorption coefficients of PFCs are relatively low for C4–C8-carboxylic acids and increase with increasing chain length (de Voogt et al. 2006). On the other hand, Tomy et al. (2004b) suggested that exposure concentrations are greater in the water column, because they observed higher concentrations in zooplankton compared to benthic invertebrates in an Arctic marine food web.

Biomagnification of PFOS in the estuarine food chain of the Western Scheldt estuary was observed by de Vos et al. (2008). The magnification ratios increase in primary and primary–secondary carnivores (carnivores that feed on both herbi-detritivores and primary carnivores), with the exception of primary–secondary carnivores to *Sterna hirundo*.

It is not clear if there is a difference between the concentrations of PFCs in edible fish from remote versus highly industrialized or urbanized areas or not. However, Gulkowska et al. (2006) observed slightly higher PFOS concentrations in fish from the highly urbanized and industrialized Guangzhou region compared to the concentration found in the more remote Zhoushan region.

In a few studies, positive correlations were found between PFC body burdens and self-reported fish consumption. In Poland, blood samples from 45 donors living near the Baltic Sea were analyzed in 2004 (Falandysz et al. 2006). Groups of people with a high consumption of regionally captured fish showed statistically higher PFC blood levels than the groups who consumed less regionally captured fish. The authors concluded that the consumption of seafood was an important determinant for internal PFC exposure. Results of another study, carried out in North Rhine-Westphalia area, also indicated a positive association between PFOS concentrations in plasma and consumption of locally caught fish, indicating that fish intake can be an important pathway for internal PFC exposures (Hölzer et al. 2008).

In recognition of the potential for human exposure to PFCs via fish consumption, the Minnesota Department of Health has recently issued fish consumption advisories for contaminated sections of the Mississippi River (Minnesota Department of Health 2007). This advisory suggests that people limit their intake of fish to no more than one meal a week, if PFOS levels in fillet exceed 38 ng/g. It is therefore interesting to note that 78.6% of the whole fish homogenates given in Table 1 had PFOS levels that exceeded that threshold. Of the muscle samples, at least five mean concentration values exceeded 38 ng/g. The relationship between measurements made with homogenates and fillets has not been examined, but the portion of all the analyzed samples that exceeded the advisory limit indicates that consumption of fish may be a route of PFC exposure that needs further evaluation.

Assuming a daily intake for fish of 50 g/d with a PFOS concentration at the advisory limit of 38 ng/g, we can calculate that humans would obtain a daily intake of 1.9 µg/d just for fish. The provisional tolerable daily intake (TDI) values proposed by the European Food Safety Authority (EFSA 2008) and Health Protection Agency (HPA 2009) amount to 150 ng/kg body weight (bwt)/d and 300 ng/kg bwt/d, respectively. For a person of 60 kg weight, this would mean a TDI of 9 and 18 µg/d, which is almost 10 times more than the estimated PFOS intake of fish (1.9 µg/d), based on the advisory limit. As this estimation is only based on fish, it is clear that the daily intake will be higher if other food and beverages are taken into account. Furthermore, the intake via non-dietary routes (e.g., dust) should also be included and would add to the total intake value.

3 Contamination of Food

Very few reports in the literature focus on concentrations of PFCs in food items. Nevertheless, it is likely that diet is an important source of human exposure to PFCs. The widespread occurrence of PFOS and PFOA in children and adults suggests that exposure results from a common source for all age groups. A wide variety of industrial and consumer applications for PFCs exist and lead to numerous possibilities for release into the environment, with subsequent human exposure via environmental routes. However, there are also more direct routes, e.g., dietary exposure, by which humans may be exposed to perfluorinated compounds (Tittlemier et al. 2007; Vestergren et al. 2008; Fromme et al. 2009; Nania et al. 2009). Vestergren et al. (2008) concluded that consumption of contaminated food and drinking water constitutes the major pathway for humans.

3.1 Indirect Contamination of PFCs in Food Items

PFCs are widely used in food-packaging coatings. Some formulations are utilized in food packaging to form grease- and water-repellent coatings for papers and paperboards. In such coatings, the mixtures of perfluorooctylsulfonyl phosphate esters used include heptadecafluoro-1-octanesulfonamide (PFOSA), N-ethyl-heptadecafluoro-1-octanesulfonamide (N-EtPFOSA) and N,N-diethyl-heptadecafluoro-1-octanesulfonamide (N,N-Et$_2$PFOSA). Perfluorooctylsulfonyl compounds may be present as manufacturing residuals in coatings and may migrate into food upon contact. Laboratory studies also indicate that some perfluorooctylsulfonyl compounds can be metabolized to PFOS (Tomy et al. 2004a; Xu et al. 2004).

Begley et al. (2005) investigated several potential sources of migration from food packaging. The most recognizable products to consumers are the uses of perfluorochemicals in non-stick coatings (polytetrafluoroethylene or PTFE) for cookware and also their use in paper coatings for oil and moisture resistance. PFOS is a residual impurity in some paper coatings used for food contact and PFOA is a processing aid in the manufacture of PTFE. Results from these authors showed that the largest

potential source of PFCs from food contact materials appears to be paper with fluorochemical coatings/additives. Migration of 4,000 mg/kg to food oil was observed
from popcorn bags. This amount of migration was hundreds of times more than
the amount of fluorochemical that was calculated to migrate at 175°C from PTFE
cookware.

Tittlemier et al. (2006) measured the concentrations of N-EtPFOSA, PFOSA;
N,N-Et$_2$PFOSA; N-methylperfluorooctane sulfonamide (N-MePFOSA), and N,N-
dimethyl perfluorooctane sulfonamide (N,N-Me$_2$PFOSA) in food items collected
over a 12-year period (1992–2004) through the Canadian Total Diet Study to estimate dietary exposure of Canadians over 12 years of age. The most frequently
detected analyte was N-EtPFOSA, found in 78 out of 151 composites, followed
by N-MePFOSA (in 25 of 51). The highest concentrations and frequency of detection of analytes occurred in the fast food composites, particularly French fries, egg
breakfast sandwiches, and pizza. Maximum concentrations of total perfluorooctane
sulfonamides analyzed in fast foods ranged from 9.7 for French fries to 27.3 ng/g
for pizza. Relatively high concentrations were also detected in cookies, Danish pas-
tries, microwave popcorn, and wieners. Data generated from the study of Tittlemier
et al. (2006) suggest that migration from food packaging had occurred.

One should realize, of course, that perfluorooctane sulfonamides can enter human
food as a result of exposure of food-producing animals or plants to these compounds
via environmental routes, such as inhalation or adsorption from air or intake of
contaminated food. Perfluorooctane sulfonamides have indeed been detected in air
(Martin et al. 2002; Stock et al. 2004; Shoeib et al. 2005) and water (Boulanger et al.
2005).

3.2 Direct Contamination of PFCs in Commercial Food Items

Tittlemier et al. (2007) analyzed 54 solid food composite samples for perfluorocarboxylates and perfluorooctane sulfonate. Forty-nine composite food samples
originated from the Canadian Total Diet Study (TDS). Just over one half of the
composites were from the 2004 TDS, the remaining composites were collected
as the TDS was organized from 1992 to 2001. The latter ones were selected for
analysis because they consisted of meat or other animal-derived food items or
could have been stored in packaging treated with grease-resistant coatings. The
food items originated from four different grocery stores and fast food restaurants.
The food items were prepared for consumption and replicate food items from the
various grocery stores or restaurants are combined and homogenized to form a
composite sample. The composite items were analyzed for PFHpA, PFOA, PFNA,
PFOS, PFDA, PFUnA, perfluorododecanoic acid (PFDoA), and perfluorotetradecanoic acid (PFTA). PFCs were detected in 9 out of 54 composites analyzed. PFOS
was the dominant compound, followed by PFOA. PFHpA and PFNA were also
detected. All the other investigated PFCs were below the detection limit, which varied between 0.4 and 5 ng/g wwt. Detected PFC concentrations ranged from 0.5 to

6 ng/g wwt. The highest PFOS concentration of 2.7 ng/g wwt was detected in beef steak, followed by a concentration of 2.6 ng/g wwt measured in a marine fish composite sample. The other PFOS concentrations ranged between 0.5 and 2.1 ng/g wwt, detected in ground beef, freshwater fish, microwave popcorn, and luncheon meats (cold cuts). Microwave popcorn contained the highest PFOA concentration of 3.6 ng/g wwt, followed by roast beef (2.6 ng/g wwt). Traceable PFOA levels were found in pizza. Only one composite contained a quantifiable PFNA concentration of 4.5 ng/g wwt. This concentration value was detected in beef steak and was the highest PFC concentration measured in this study. In two composites (pizza and microwave popcorn), low concentrations of PFHpA were found.

Researchers, in a study carried out by the UK Food Standards Agency (FSA 2006), analyzed 20 composites from the 2004 TDS for PFCs. Detectable levels of PFOS were found in canned vegetables, eggs and sugars, and in preserved food groups. PFOA was only detected in the potatoes group which included potato chips, French fries, instant mash, and other potato products. These results may have been due to analytical artifacts (Mortimer DN, personal communication). Other fluorinated chemicals PFOSA, perfluorobutane sulfonate (PFBS), PFHxS, PFHxA, PFNA, PFDA, PFUnA, PFDoA, perfluorotetradecanoic acid (PFTA) were detected only occasionally, although 10 different fluorinated compounds were found in the potatoes food group. PFHpA, perfluorohexadecanoic acid (PFHdA), and perfluorooctadecanoic acid (PFOdA) were not found. Bread, cereals, carcass meats, offal, meat products, poultry, fish, green vegetables, fresh fruit, beverages, milk, and nuts did not contain quantifiable concentrations of PFCs.

In a study conducted for 3 M Environmental Technology and Safety Services by the Battelle Memorial Institute (Columbus, OH, USA) (US EPA 2001), preliminary data about the presence of fluorochemicals in foods and in drinking water were collected and analyzed on PFOS, PFOA, and perfluorooctane sulfonamide (PFOSA). Each of three cities having manufacturing or commercial use of fluorochemical products (test cities) were paired with three cities that did not (control cities). A total of 12 samples were found to contain levels of PFCs above the limit of quantification. Of the 12 samples with measurable fluorochemical residue levels, 8 were samples collected in test cities. Measurable quantities of PFOS were found in five samples: four whole milk samples (three from test cities) and a ground beef sample (test city). PFOS residues found in the food samples ranged from non-quantifiable levels to 0.852 ng/g wwt. Measurable quantities of PFOA were found in seven samples: two ground beef samples (neither from test cities); two bread samples (one from a test city); two apple samples (both from test cities); and one green bean sample (from a test city). PFOA residue levels ranged from non-quantifiable levels to 2.35 ng/g wwt. A value of 14.7 ng/g wwt was found for PFOA in a bread sample from a control city, but was considered "suspect." An important remark on these results is that only for the ground beef sample were these concentrations found in the replicates. The remaining values above the quantification limit were not detected in the duplicates. In general, the distributions of the PFOS, PFOA, and PFOSA residue data by food and city category reveal similar patterns of residue concentrations in the control and test cities for each type of food.

Ericson et al. (2008a) determined PFC levels in 36 composite samples of food-stuffs randomly purchased in various locations to determine the dietary intake of PFCs by the population of Tarragona County (Spain). PFOS, PFOA, and PFHpA were the only detected PFCs in the composite food samples. Fish, followed by dairy products and meats, were the main contributors to PFOS intake. The authors esti-mated the exposure to PFCs through the diet for various age and gender groups. Their results did not justify dietary intake as being the main route of exposure governing blood concentrations of other PFCs (Ericson et al. 2008a).

Fromme et al. (2007b) quantified the dietary intake of PFOS, PFOA, PFHxS, PFHxA, and PFOSA using 214 duplicate diet samples for an adult study popula-tion in Germany. The participants collected daily duplicate diet samples over seven consecutive days in 2005.

Overall, they detected PFOS, PFOA, PFHxS, and PFHxA in 33, 45, 3, and 9% of the 214 single duplicate samples, with concentrations ranging from <LOD (limit of detection) to 1.03 ng/g wwt, <LOD to 118.3 ng/g wwt, <LOD to 3.03 ng/g wwt, and <LOD to 3.2 ng/g wwt, respectively. PFOSA could not be detected above the LOD. Including only values above the detection limit ($n = 47$), a significant correla-tion was observed between PFOS and PFOA concentrations. The calculated median daily intake of PFOS and PFOA was 1.4 ng/kg bwt and 2.9 ng/kg, respectively. The median daily intake of PFHxS and PFHxA was 2.0 and 4.3 ng/kg bwt, respectively. The authors remarked that for the interpretation of this data, it has to be kept in mind that these analytes were detected only in few samples. In this study, gender differences were not observed for the target analytes.

Kärrman et al. (2009) investigated the relationship between dietary exposure and serum PFC levels in two Japanese cities. Therefore, 1-day composite diet samples from 20 women in Japan were collected in 2004 and were analyzed for PFBS, PFHxS, PFOS, PFHxA, PFHpA, PFOA, PFNA, PFDA, and PFUnA. This was done in a remote area and in an urban area so that the possible influence of industrial-ization could be investigated. Only PFOS and PFOA could be detected in the diet samples (within a range of 0.008–0.087 and 0.008–0.040 ng/g fresh wt, respec-tively). The levels of PFOA were relatively close to the method detection limit. PFOS and PFOA were detected in all diet samples and no difference was observed between the remote and the urban area. However, the importance of the diet for the body burden seems to vary between regions. The authors concluded that the dietary intake of PFOS and PFOA accounted for 22.4 and 23.7% of serum levels in females from the urban area and, in contrast, for 92.5 and 110.6% in females from the non-industrialized area, respectively.

Dietary exposure to PFCs has also been indirectly examined by Falandysz et al. (2006), who observed a correlation between PFC concentrations in blood sampled from adults in Poland and self-reported consumption of Baltic fish. However, no estimation of dietary exposure to PFCs could be made from this study.

Concentrations observed in all the studies conducted to date were in the sub- to low-ng/g range. It is worth noting that concentrations of PFCs in individual food items used to prepare the composite samples will be higher than those reported for the composite ones, since PFC-free food items in the same composite can effectively dilute PFC concentrations in individual food items (Tittlemier et al. 2007). Cooking practices may also lead to additional contamination of the prepared meal, e.g., by transfer of PFCs from frying pans or from contaminated water used for cooking (Begley et al. 2005). On the other hand, Del Gobbo et al. (2008) suggested that cooking methods (baking, boiling, and frying) can reduce PFC concentrations in fish.

The PFC levels found in food reflect both the environmental exposure and the food-packaging sources of entry into prepared foods, since both animal-derived and vegetable-based foods were found to contain PFCs.

4 PFCs in Drinking Water

In this section, an overview of existing data on PFCs in drinking or tap water is given. To the best of our knowledge, studies on other drinks like soda, coffee, tea, or juices have not been reported in the literature.

Results for PFCs found in drinking water are summarized in Table 4. Fromme et al. (2009) summarized existing data on PFCs in drinking water. In potable tap water of Japan, PFOS concentrations between 0.1 and 51 ng/L were detected. The majority of the results for individual samples did not exceed 4 ng/L (Harada et al. 2003). Saito et al. (2004) reported concentrations of PFCs in drinking water from areas with known PFC sources; their results ranged between 5.4 and 40.0 ng/L for PFOS. For PFOA, the concentrations ranged between 1.1 and 1.6 ng/L, while in areas with no known sources, concentrations ranged from <0.1 to 0.2 ng/L for PFOS and from 0.1 to 0.7 ng/L for PFOA.

In a study in which tap water that originated from Italy's Lake Maggiore was examined (Loos et al. 2007), all investigated PFCs (PFHpA, PFOA, PFOS, PFNA, PFDA, PFUnA, and PFDoA) were detected, with concentrations ranging from 0.1 to 9.7 ng/L. The highest values were 9.7 ng/L for PFOS, 2.9 ng/L for PFOA, and 2.8 ng/L for PFDoA.

Skutlarek et al. (2006) analyzed drinking water samples in a contaminated area of the Ruhr and Moehne catchment (Germany). The highest values found were at Neheim, and the PFOA levels found were 519 ng/L, followed by PFHpA (23 ng/L) and PFHxA (22 ng/L). Measured PFC levels in the tap water ($\sum$PFC = 767 ng/L) were comparable to the levels in water from the Moehne river ($\sum$PFC = 598 ng/L) that served as a water supply. The sum of PFCs in drinking water from Ruhr waterway supplies ranged between 54 and 301 ng/L. In drinking water from the northern part of Duisburg, all the PFCs were below the LOD. In the southern district of Duisburg, however, PFBS was determined at 26 ng/L. The observed variation in concentrations in Duisburg indicates that samples were collected from different water supplies. Perfluorinated chemicals at other selected drinking water sampling sites outside the Ruhr and Moehne were sporadically detected, with the highest observed concentration of PFBS in Koblenz (20 ng/L). The sum of PFCs for those sites varied between non-detected and 27 ng/L (Skutlarek et al. 2006).

Table 4 Observed range of perfluorinated chemicals in drinking water (ng/L)

Sampling sites (n) country, sampling date	PFHxA*	PFHpA	PFOA	PFBS	PFHxS	PFOS	PFNA	PFDA	PFUnA	PFDoA	References
Tokyo, Kyoto Japan, not given						0.1–51					Harada et al. (2003)
Osaka and Tohoku, area, Japan 2003			0.7–50			<0.1–12					Saito et al. (2004)
Rhine, Ruhr Moehne area, Germany 2006	<1–56	<1–23	<1–519	<1–26		<1–22					Skutlarek et al. (2006)[a]
Area of Lake Maggiore (n = 6) Italy 2007		0.3–0.8	1.0–2.9			6.2–9.7	0.3–0.7	0.1–0.3	0.1–0.4	0.1–2.8	Loos et al. (2007)
Catalonia (n = 4) Spain 2007	<0.9	<0.6–3.0	0.3–6.3	<0.3	<0.2–0.3	0.4–0.9	<0.4–0.5	<0.8	<0.4	<0.3	Ericson et al. (2008b)[b]
Catalonia (n = 4)[c] Spain 2007	<0.9	<0.6–0.4	<0.2–0.7	<0.3	<0.2	<0.24	<0.4–0.2	<0.8–0.6	<0.4	<0.3	Ericson et al. (2008b)[b]
Osaka (n = 14) Japan 2006–2007			2.3–84			<0.1–22					Takagi et al. (2008)

*PFHxA: perfluorohexanoic acid; PFHpA: perfluoroheptanoic acid; PFOA: perfluorooctanoic acid; PFBS: perfluorobutane sulfonate; PFHxS: perfluorohexane sulfonate; PFOS: perfluorooctane sulfonate; PFNA: perfluorononanoic acid; PFDA: perfluorodecanoic acid; PFUnA: perfluoroundecanoic acid; PFDoA: perfluorododecanoic acid

[a] Skutlarek et al. (2006) measured also PFBA (perfluorobutanoic acid, <1–11); PFPeA (perfluoropentanoic acid, <1–77)

[b] Ericson et al. (2008b) measured also PFBS (<0.3); THPFOS (1,1,2,2,-tetrahydro perfluorooctanoic acid, <1.0); PFOSA (<0.19); PFDS (perfluorodecane sulfonate, <1.0), and PFTdA (perfluorotetradecanoid acid, <0.9)

[c] Bottled water

Ericson et al. (2008b) analyzed 14 PFCs in drinking water (tap and bottled) from Tarragona Province (Catalonia, Spain). In 2007, municipal drinking (tap) water was obtained in public fountains of the three most populated towns in Tarragona Province. The bottled water samples from four commercial companies, whose water spring has different origins, were purchased from a supermarket. This is, to the best of our knowledge, the first study in which bottled water was analyzed. The PFC levels in tap water varied among the four places. In the Valls sample, the highest PFC levels were found; PFHpA (3.02 ng/L), PFOS (0.44 ng/L), and PFOA (6.28 ng/L). In the sample of tap water from Reus, PFHxS (0.28 ng/L), PFOA (0.98 ng/L), PFNA (0.52 ng/L), and PFOS (0.73 ng/L) were detected, while in that of Tarragona, PFHpA (0.64 ng/L) PFHxS (0.28 ng/L), PFOA (0.98 ng/L), PFNA (0.22 ng/L), and PFOS (0.87 ng/L) were found. The lowest contamination in tap water corresponded to the sample collected in Tortosa, in which only PFOA (0.32 ng/L) and PFOS (0.39 ng/L) were detected. The samples of bottled water contained some PFCs at levels that corresponded to the lowest values observed in tap water. In one sample (Veri) all the PFCs were below the respective limits of detections, whereas in those of Cardo and Caprabo only PFNA (0.13 ng/L) and PFOA (0.67 ng/L) could be detected. In the fourth sample (the Font Vella) PFHpA (0.40 ng/L), PFOA (0.34 ng/L), and PFNA (0.20 ng/L) were detected. Overall, the PFC levels were notably lower in bottled water, whereas PFOS could not be detected in any sample. The authors also determined the levels of daily intake of PFOS through tap water, assuming an intake of 2 L tap water/d for the four different sampling towns. The result varied between 0.78 and 1.74 ng/d (absolute). Ericson et al. (2008b) estimated, in a previous study, a dietary intake of PFOS at 62.5 or 74.2 ng/d (assuming ND = 0 or ND = $\frac{1}{2}$ LOD, respectively) for an adult of 70 kg bwt living in the Tarragona Province.

In 2006–2007, tap water was collected from 14 water treatment plants in Osaka, one of the largest industrial cities of Japan. PFOS was detected in 25 tap water samples and PFOA was detected in all of the tap water samples analyzed. Concentrations of PFOS and PFOA in tap water were 0.16–22 and 2.3–84 ng/L, respectively (Takagi et al. 2008).

Dutch drinking water resources are known to contain PFCs, from non-detect to 43 ng/L (Mons et al. 2007), and little is known about the behavior of these compounds in the Dutch drinking water preparation cycle.

The Bureau of Safe Drinking Water (BSDW) initiated a preliminary occurrence study in July 2006 to determine whether PFOS and PFOA could be found in detectable concentrations in raw and treated public water systems throughout the state of New Jersey (USA). Five out of 23 public water samples showed non-detectable levels of PFOA. Detected and quantifiable levels were found in 15 samples with values ranged from 4.5 to 39 ng/L. Additionally, three samples showed levels of PFOA that were detected but not quantified (<4 ng/L). Ten out of 23 water samples collected from public water systems showed non-detectable levels of PFOS. In seven samples PFOS concentration could be quantified and varied between 4.2 and 19 ng/L. Six samples showed levels of PFOS but could not be quantified (<4 ng/L) (New Jersey Department of Environmental Protection Division of Water Supply 2007).

Hölzer et al. (2008) analyzed drinking water (tap) to investigate the correlation between the PFC concentration in water and serum of the local people in Arnsberg (Germany). As previously mentioned, Skutlarek et al. (2006) observed remarkably high PFC concentrations in surface waters in this area (Moehne and Ruhr catchment). Tap water samples were collected from the kitchen in the homes of all residents. Of the various analyzed PFCs, PFOA was the main compound found in drinking water (500–640 ng/L). In the drinking water samples from reference areas, Siegen and Brilon, PFOS and PFOA were not detected. After installation of activated charcoal filters in the waterworks, PFOA concentrations in Arnsberg were significantly reduced. However, during the study period, filtration performance declined and PFOA concentrations in tap water samples increased from below the LOD to 71 ng/L. As the detected blood concentrations of residents living in Arnsberg were 4.5–8.3 times higher than those for the reference populations, the authors concluded that the consumption of tap water at home was a significant predictor of PFOA blood concentrations in Arnsberg.

5 Safety Limits and Tolerable Daily Intakes

It must be noted that the international regulatory organizations (World Health Organization (WHO), European Union (EU)/EFSA, US EPA, etc.) have not established safety limits yet for PFCs in drinking water. However, for guidance purposes, the 3 M Company developed a lifetime Drinking Water Health Advisory (DWHA), which was estimated to be 0.1 μg/PFOS/L (assuming a consumption of 2 L/day of contaminated water; 3 M 2001). Recently, Schriks et al. (2010) derived provisional drinking water guideline values for PFOS and PFOA of 0.5 and 5.3 μg/L, respectively, on the basis of the TDI values proposed by EFSA (2008).

At a PFC works facility, near Washington, West Virginia, the action level of PFOA in drinking water agreed to between the US EPA and E.I Dupont de Nemours was 0.50 μg/L. If the PFOA level in water supplies reaches 0.5 μg/L the local producer of PFCs must take certain actions (such as providing residents with bottled water). When the PFOA level in drinking water is measured at 150 μg/L, the producer has to install water treatment equipment (http://www.epa.gov/region03//enforcement/dupont_factsheet.html). In 2006, EPA and the eight major PFC manufacturing companies in the industry launched the 2010/15 PFOA Stewardship Program, in which companies committed to reduce global facility emissions and product content of PFOA and related chemicals by 95% by 2010 and to work toward eliminating emissions and product content by 2015 (http://www.epa.gov/oppt/pfoa/pubs/stewardship/index.html).

Recently, in New Jersey, the Department of Environmental Protection developed preliminary health-based drinking water guidance for PFOA of 40 ng/L (http://www.defendinscience.org/case_studies/upload/pfoa_dwguidance.pdf).

Guidelines have also been developed in Europe, more specifically in Germany. The occurrence of PFCs in surface and drinking waters of the Ruhr and Moehne area (Skutlarek et al. 2006) caused considerable concern, in view of the possible effects on humans and the ecosphere. Therefore, German authorities have recommended guide values for PFOA and PFOS in drinking water. Immediately after the increased PFOA levels were observed, German Drinking Water Commission (DWC) of the German Ministry of Health at the Federal Environment Agency established guide values for human health protection. Precautionary actions for infants will be taken if PFOA and PFOS concentrations in drinking water reach 0.5 µg/L. If additional PFCs are present, the health-based precautionary value (long-term minimum quality goal) becomes 0.1 µg/L. As a result, local health authorities recommended that residents in parts of Arnsberg to not use the drinking water for preparation of baby food and advised pregnant women to avoid regular intake of such water (http://www.umweltbundesamt.de/uba-info-presse-e/hintergrund/pft-in-drinking-water.pdf).

Most monitoring studies have focused only on PFOS and PFOA, but a few also reported on other PFCs that appear at rather high concentrations in potable water such as PFBS, PFDoA, perfluoropentanoic acid (PFPeA), and PFHxA (Skutlarek et al. 2006; Loos et al. 2007, Ericson et al. 2008b). Therefore, it is important to increase monitoring efforts with a view to setting more comprehensive safety limits for PFCs in potable water.

The relatively high concentrations of PFCs that have been observed in drinking water samples indicate that the common water treatment steps used do not effectively eliminate perfluorinated compounds. It should be noted that the washing of food samples with tap water may introduce an additional source of PFCs (Ericson et al. 2008b).

Recently, several scientific institutions have derived TDIs from toxicological end points by applying an uncertainty factor. The Scientific Panel on Contaminants in the Food Chain (CONTAM) established a TDI for PFOS of 150 ng/kg bwt/d and for PFOA of 1.5 µg/kg bwt/d (EFSA 2008). The UK Committee on Toxicity of Chemicals in Food, Consumer Products and the Environment (COT) proposed a TDI for PFOS and PFOA of, respectively, 300 and 3,000 ng/kg bwt/d (COT 2006a, b). Furthermore, the German Federal Institute for Risk Assessment proposed a TDI of 100 ng/kg bwt/d for both PFOS and PFOA (BfR 2006).

These intake values are notably higher than those derived from actual human exposure via diet and beverages. For example, Fromme et al. (2007b) estimated a median dietary intake of 1.4 ng PFOS/kg bwt/d and of 2.9 ng/PFOA kg bwt/d. The results were similar to those calculated in the study of Kärrman et al. (2009) for two regions in Japan. They calculated a median daily intake of PFOS and PFOA, respectively, from 1.1 to 1.5 ng/kg bwt/d and from 0.72 to 1.3 ng/g bwt/d. In other studies, estimates of the daily intake were derived through market-basket surveys of food items instead of duplicate diet samples. In Canada, the average dietary intake of total PFCs and PFOS for individuals between the ages of 12 and those over 65 was estimated to be 250 ng/d (Tittlemier et al. 2007). This estimate was based upon results from 25 food composite samples, which did not represent the whole diet.

These results represented only food that could have been environmentally exposed through bioaccumulation of PFCs or may have come into contact with food packaging treated with PFCs. The UK Food Standards Agency calculates average dietary intakes from the whole diet; these calculations give intake values of 100 ng/kg bwt/d for PFOS and 70 ng/kg bwt/d for PFOA. The UK dietary intake estimate is notably higher than that derived from other studies, probably due to the relatively high concentrations found in the potato composite samples. Finally, a food market study conducted in Spain (Tarragona County) resulted in a PFOS daily intake estimate of 1.07 ng/kg bwt/d for adult men (Ericson et al. 2008a). The EFSA (2008) study estimates indicated actual daily intakes of 60 ng/kg bwt/d of PFOS and 2 ng/kg bwt/d of PFOA.

Although diet and drinking water are thought to be two of the major exposure sources for humans, one should remember that all of these daily intake values, including the (provisional) TDIs cited above, are derived solely from food and/or beverage PFC levels (mainly PFOS and PFOA). Non-dietary exposure routes were not included and neither were the contributions of precursor compounds. As mentioned before, the calculated daily intake values are lower than the TDIs proposed. Nevertheless, these estimates were calculated for adults and the intake for infants, toddlers, and children may be much higher. As Trudel et al. (2008) have shown, the youngest among the consumer groups tend to experience higher total intake doses (on a body weight basis) than do teenagers and adults; this higher uptake results from the higher relative intake via food consumption and also hand-to-mouth transfer of PFCs from treated carpets and ingestion of dust.

6 Perfluorinated Compounds in House Dust and Air

In a few studies, the presence of perfluorinated compounds has been documented to occur in the indoor environment (Table 5). Given the wide range of consumer and residential products that contain or have been treated with PFCs, it is reasonable to hypothesize that as these products age and degrade their debris will accumulate indoors. Perhaps the most obvious potential source of PFCs in indoor dust would be the anti-stain agents used on carpets and upholstery (Strynar and Lindstorm 2008).

Moriwaki et al. (2003) analyzed dust for PFOS and PFOA in 16 Japanese houses. They collected bags from vacuum cleaners that were voluntarily donated. Hair and plastic garbage were removed from the samples using forceps and a loupe. In each sample, both analytes were found at levels above the limit of determination. The concentrations ranged from 11 to 2,500 ng/g for PFOS and from 69 to 3,700 ng/g for PFOA. In one sample, extremely high concentrations of 2,500 (PFOS) and 3,700 (PFOA) ng/g were found. Without this sample, the maximum concentrations were 120 ng/g for PFOS and 380 ng/g for PFOA. For all the dust samples analyzed in this study, the concentrations of PFOA were higher than those for PFOS (Moriwaki et al. 2003).

Table 5 Median, minimum, and maximum concentrations (ng/g) of PFOS, PFOA, and PFHxS detected in indoor house dust

References, sample location, (n)		PFOS	PFOA	PFHxS
Moriwaki et al. (2003)	Median	24.5	165	/
Japan (16)	Min	11	69	/
	Max	2,500	3,700	/
Kubwabo et al. (2005)	Median	37.8	19.7	23.1
Canada (67)	Min	2.3	1.2	2.3
	Max	5,065	1,234	4,305
Nakata et al. (2007)	Median			/
Japan (7)	Min	7	18	/
	Max	41	89	/
Strynar and Lindstorm	Median	201	142	45.5
(2008) North America	Min	<8.9	<10.2	12.9
(102, 10[a])	Max	12,100	1,960	35,700
Fromme et al. (2008)	Median	16	11	/
Germany (12)	Min	3	2	/
	Max	342	141	/
Björklund et al. (2009)	Median	39	54	/
Sweden (10 houses)	Min	15	15	/
	Max	120	98	/
Björklund et al. (2009)	Median	85	93	/
Sweden (38 apartments)	Min	8[b]	17	/
	Max	1,100	850	/
Björklund et al. (2009)	Median	110	70	/
Sweden (10 offices)	Min	29	14	/
	Max	490	510	/
Björklund et al. (2009)	Median	12	33	/
Sweden (5 cars)	Min	8[b]	12	/
	Max	33	96	/
Goosey et al. 2008	Median	1,200	220	/
UK (unknown)[c]	Min	85	42	/
	Max	3,700	640	/

[a]Day-care centers
[b]Limit of quantitation/2
[c]Primary school and nursery classrooms; /: not analyzed

In dust samples (particle size of 75 μm to 1 mm) from seven Japanese houses, PFOA and PFOS were detected; their levels ranged from 18 to 89 ng/g and 7 to 41 ng/g, respectively (Nakata et al. 2007).

In Canada, dust samples were investigated for PFCs from 67 randomly selected homes during the winter of 2002/2003 (Kubwabo et al. 2005). The dust was collected from vacuum cleaners and sieved (opening 1.18 mm) before analysis of the samples for PFBS, PFHxS, PFOA, PFOS, and PFOSA. Results indicate that PFOA, PFHxS, and PFOS had much higher detection frequencies (63, 85, and 67%, respectively) than did PFOSA, which was detected in 10% of the samples. PFBS was not detected in any of the samples, possibly because this compound had just been introduced to the market at the time of these analyses. The concentrations of PFOA, PFOS, and PFHxS varied between 1.2 and 1,234, between 2.4 and 5,065, and between 2.3 and 4,305 ng/g, respectively.

In 2000–2001, dust samples from 102 homes and 10 day-care centers in North Carolina and Ohio were collected (Strynar and Lindstrom 2008). The samples were taken from vacuum cleaning bags and sieved to remove particles of diameters >150 μm. Quantifiable levels of PFOA, PFOS, and PFHxA were detected in 96.4; 94.6, and 92.9% of the samples, respectively. This is, to our knowledge, the first study in which PFHxA, PFHpA, PFNA, PFDA, PFUnA, and PFDoA were analyzed in dust samples. These analytes were, respectively, detected in 93, 74, 43, 30, 37, and 19% of the analyzed samples. PFBS was detected in 33% of the samples. The highest concentration observed among the samples analyzed amounted to 35,700 ng/g for PFOS. Generally, the fluorotelomer alcohols or FTOHs (6:2, 8:2, and 10:2 FTOHs) were detected in 50% of the samples and had median concentrations that ranged between 24 and 33 ng/g (Strynar and Lindstrom 2008).

Additionally, in a pilot study in Germany, 12 dust samples were collected using a vacuum cleaner. Median (range) PFOS and PFOA concentrations in the sieved fraction (particles < 630 μm) were 0.016 and 0.011 μg/g, respectively. Significantly lower median concentrations were observed in the unsieved samples, indicating that PFCs were mainly associated with smaller particles (Fromme et al. 2008).

Shoeib et al. (2005) were the first to analyze N-methylperfluorooctane sulfonamidoethanol (MeFOSE), N-ethylperfluorooctane sulfonamidoethanol (EtFOSE), N-ethylperfluorooctane sulfonamide (EtFOSA), and N-methylperfluorooctane sulfonamidethylacrylate (MeFOSEA) in indoor dust samples. During the winter of 2002/2003, dust was collected from 66 randomly selected homes in the city of Ottawa, Canada. The dust used for analysis came from the occupants' vacuum bags or central vacuum containers. MeFOSEA was detected in ~30% of dust samples and had a geometric mean value of ~8 ng/g. The highest concentrations in dust were observed for MeFOSE and EtFOSE, which had geometric means of 113 and 138 ng/g, respectively (Shoeib et al. 2005).

During the winter of 2007 and spring of 2008, dust samples were collected from primary school and nursery classrooms in the West Midlands of the UK using a portable vacuum cleaner to which a sock with a 25-μm mesh size was inserted. After filtration through a 500-μm mesh the samples were analyzed for PFOA and PFOS. The concentrations of PFOA and PFOS found varied between 42 and 640 and between 85 and 3,700 ng/g, respectively (Goosey et al. 2008). Recently, a study was published in which the dust samples collected in Stockholm city (Sweden) from 10 houses, 38 apartments, 10 day-care centers, 10 offices, and 5 cars were investigated (Björklund et al. 2009). The dust samples were collected during 2006/2007 on pre-weighed cellulose filters in styrene–acrylonitrile holders that were inserted into a polypropylene nozzle; the nozzle was then attached to the intake nozzle of an industrial strength vacuum cleaner. Sampling of surfaces was performed at least one meter above the floor; sampled surfaces included bookshelves, moldings, counters, and lampshades. For cars, plastic surfaces and seat covers were vacuumed. Analysis of the samples from the 38 apartments disclosed PFOS and PFOA residues in 79 and 100% of the samples, respectively. PFOS was detected in 3 out of 5 car samples, whereas PFOA was detected in all of the car samples. PFOA concentrations were higher than those of PFOS in 60% of all the samples. Although offices had the highest median PFOS concentrations, the highest individual concentrations were found in apartments. The highest individual PFOA concentrations were also found in some apartments. The highest variation for both compounds was seen in apartments. Houses and day-care centers had much lower concentration variability. The median concentrations found in the different microenvironments were within one order of magnitude of each other. The highest median PFOS concentrations were seen in offices (110 ng/g dwt); similar but lower concentrations were seen in apartments (85 ng/g dwt), houses (39 ng/g dwt), and day-care centers (31 ng/g dwt); and lowest concentrations were seen in cars (12 ng/g dwt). For PFOA, the concentrations were more similar than was PFOS, between different microenvironments, with highest median concentrations found in apartments (93 ng/g dwt) and offices (70 ng/g dwt). Offices had higher median PFOS than PFOA concentrations, whereas the opposite was found for the other microenvironments (Björklund et al. 2009).

In Norway, PFOS levels were found in May 2005 in the particulate phase of one indoor air location. The concentrations were below the limit of determination ($=$ 0.0474 ng/m^3). For PFOA levels ranged between 0.0034 and 0.0069 ng/m^3 (Barber et al. 2007).

Concentrations of PFOS and PFOA in the particulate phase of European air were also reported (Barber et al. 2007). Air samples were collected in 2005 at two locations in the UK and two locations in Norway. PFOS levels in an urban area ranged from 0.041 to 0.051 ng/m^3 in the UK in March and from 0.0009 to 0.0071 ng/m^3 in the UK in November. In southern Norway, PFOS levels ranged from 0.0009 to 0.0011 ng/m^3 in November. The levels of PFOS in the particulate phase of air in the UK were the highest reported anywhere to date. Levels of PFOA in the particulate phase of UK air varied from 0.226 to 0.828 ng/m^3 in March 2005, and from 0.006 to 0.222 ng/m^3 in November. Differences in PFOA levels between the rural and the urban site in the UK were less clear than for PFOS. The levels of PFOA at the rural Norwegian site were significantly lower than those found in the UK. In southern Norway (data from November), levels varied between 0.0014 and 0.0017 ng/m^3.

7 Correlation Between PFCs

In some studies, the correlation between the analyzed PFCs was investigated to determine any potential associations. A significant positive relationship between the concentrations of PFOS and PFOA was observed in dust from Japanese houses (Moriwaki et al. 2003). This relationship was confirmed by Kubwabo et al. (2005), Goosey et al. (2008), and Björklund et al. (2009). The significant correlation between PFOS and PFOA, observed in these studies, suggests that these compounds may be from a common source or may originate from the same precursor compound.

Kubwabo et al. (2005) observed a significant and positive relation between PFOS, PFOA, and PFHxS in the dust samples of homes in Ottawa. A strong correlation between PFOS and PFHxS (Spearman's rank-order correlation coefficient (r_s) of 0.868, $p < 0.0001$) suggests that these compounds originated from the same sources in house dust (Kubwabo et al. 2005).

In addition, Strynar and Lindstorm (2008) observed that, generally, almost all of the investigated compounds (PFHxA, PFHpA, PFOA, PFNA, PFDA, PFUnA, PFDoA, PFOS, and PFHxS) had significant correlations with each other except PFBS. The FTOHs, detected in the dust samples, were highly correlated with each other ($r_s > 0.82$, $p < 0.0001$) suggesting a common source of these compounds (Strynar and Lindstorm 2008). No significant correlation between the perfluorinated sulfonamides, MeFOSE, and EtFOSE was observed (Shoeib et al. 2005). Kubwabo et al. (2005) also investigated the relationship between the analyzed PFCs and the characteristics of the house. Their results indicated that lower levels of PFOA and PFOS were found in older homes. Furthermore, PFOA, PFOS, and PFHxS and the total of these PFCs were significantly and positively correlated with the percentage of carpeting found in the house. The authors noticed that in addition to carpet, other multiple sources of perfluorinated compounds in homes may exist (Kubwabo et al. 2005).

Generally, in the Japanese studies, the PFOA concentrations were higher than those for PFOS (Moriwaki et al. 2003; Nakata et al. 2007). A similar trend was observed in Sweden, except for the levels found in offices (Björklund et al. 2009). The opposite was observed for the studies carried out in the USA (Strynar and Lindstorm 2008), Canada (Kubwabo et al. 2005), UK, and Germany (Fromme et al. 2008), where the PFOS concentrations were higher than PFOA. This observation could suggest a difference in PFC sources. The median PFOS concentration measured in the American study were 5–12 times higher than in the other studies. The median PFOA concentration in the USA was roughly comparable with one study in Japan and also 7–13 times higher than the median concentration found in the other Japanese study. Nevertheless, the ranges were similar between the studies except for the studies from Nakata et al. (2007) and Fromme et al. (2008), who measured much lower maximum values. A possible reason for the observed differences is that Strynar and Lindstorm (2008) included dust from day-care centers, where possibly, more PFC-containing waxes and cleaning products were used. Possibly, this can also explain the high median concentrations of PFOS (1,200 ng/g) and PFOA (220 ng/g) that were detected in the primary schools and nursery classrooms in the UK. More data are necessary to draw conclusions on the possible differences between geographic regions. Further research is also required to understand the different parameters contributing to the presence of perfluorinated compounds in house dust and in the indoor environment, in general (Strynar and Lindstorm 2008).

The concentrations summarized in Table 5 indicate that dust is a potential reservoir of different PFCs in homes, and these may be available for human exposures. Given that human exposure routes remain poorly characterized, the potential role of house dust needs to be more completely evaluated (Strynar and Lindstorm 2008). Additional studies are also needed to address the correlation between the levels of

PFCs in blood of humans occupying the house being tested and the concentration of PFCs in house dust (Kubwabo et al. 2005).

8 Outlook

From the discussion above it is obvious that it is imperative to further assess the origin of PFCs in the human diet and the diet's contribution to total human PFC exposure. Several initiatives have been taken recently to that end; these include the development of more robust and reliable analytical tools for the determination of PFCs, which will enable the qualification and quantification of PFCs in our diet, and the understanding of how PFCs are transferred from the environment into dietary items. In Europe, several recently launched scientific projects (e.g., EU projects Perfood; Confidence; Safefoodera; Norman network, and some national programs, e.g., in Norway, UK, and Belgium) deal with or emphasize the human exposure to PFCs and related compounds. These projects involve method development, source apportionment, exposure modeling, risk assessment, migration from packaging materials and food preparation methods, etc. Newly gained knowledge from these completed projects will enable us to evaluate the possible routes, including their relative importance, of human exposure to PFCs via our diet, to assess the role of the technosphere in the contamination of our food, help identify ways to reduce PFC contamination of dietary articles, and will help to establish relevant safe guideline values for dietary items.

Meanwhile, the steady, almost exponential increase of literature published illustrates the huge efforts taken to elucidate PFOS levels in the environment and the possible effects of PFCs and their mechanisms of action. The awareness of the current PFC issue by all stakeholders is reflected in joint efforts, such as the Stewardship Program in the USA, cited above, and joint initiatives in Europe between academic researchers and industry (e.g., PERFORCE 1 and 2 projects (see www.science.uva.nl/perforce)). The European Union, USA, Canada, Japan, UK, Norway, Sweden, and Australia, as well as several international groups, such as the OECD, are developing strategies to reduce PFC emissions and find safer alternatives to their use. Bans on use of PFOS in consumer products have been issued (EU 2006) and PFOS has been listed as a POP under the Stockholm Convention. The next decade of investigations will tell us whether or not the initiatives taken will lead to a substantial decrease in environmental and dietary exposure levels and if the recent reports of decreasing levels observed in human blood and in wildlife are a sign that the measures taken are indeed effective.

9 Summary

The widespread distribution and the degradation of PFCs in the environment results in a very complex exposure pattern, which makes it difficult to define the relative contribution to human exposure from different exposure pathways. The present review is designed to provide an overview of the existing data on levels of PFCs measured in the human diet and in drinking water. Data on levels of PFCs in the human diet are rather scarce, but the levels in fish appear to be well documented. Among PFCs, PFOS and PFOA are the best studied compounds in fish from both experimental and monitoring studies. Recently, the number of publications that address other PFCs has increased, but the total number available is still limited. In general, we discovered that care should be exercised when using the reviewed data, because, in the majority of the publications, quality control and/or details on analyses are, at least partly, lacking.

It has been well documented that PFOA and PFOS have the potential to accumulate in fish and concentrations up to 7 and 170 ng/g wwt, respectively, in edible fish species have been found. PFOS is the most crucial and prominent compound identified, followed by PFOA. Also, in aquatic invertebrates such as shrimps, mussels, clams, and oysters, high PFOS values have been reported (up to 387 ng/g wwt). However, in most publications PFC levels reported in molluscs were less than 1 ng/g wwt. Positive correlations were found between PFC body burdens and self-reported fish consumption. In recognition of the potential for human exposure to PFCs via fish consumption, the Minnesota Department of Health has recently issued fish consumption advisories for contaminated sections of the Mississippi River. It is interesting to note that 79% of the reviewed publications on PFCs in whole fish homogenates exceed that threshold. Moreover, five of the PFC concentrations reported in muscle tissue exceeded the advisory level of 38 ng/g wwt. Even though several authors concluded that consumption of contaminated food and drinking water constitutes the major exposure pathway for humans, only a few reports on PFCs in composite food exist. Food can be contaminated in an indirect way, because PFCs are widely used in food-packaging coatings and cooking materials. On the other hand, PFCs can also enter food organisms via environmental routes such as inhalation or adsorption from air. In a few studies, composite samples, duplicate diet samples, or other food items were analyzed for several PFCs. PFOS, PFOA, PFHpA, PFHxA, and PFHxS were measured and displayed concentrations ranging from non-detected up to 15 ng/g wwt. In one study, a very high PFOA concentration of 118 ng/g wwt was reported, but overall, PFC levels are below 10 ng/g wwt. It is important to note that, among all studies reviewed, PFCs were found in a maximum of 50% of the analyzed samples and generally only in 10% or less of samples analyzed.

In contrast to what is observed in fish and other food items, PFOA levels in drinking water (ND – 50 ng/L) usually exceed levels of PFOS (ND – 51 ng/L) and other PFCs (1–3 ng/L). In one study, extremely high values (519 ng/L) were measured in drinking water of a contaminated area in the Ruhr region. In Spain, bottled water was analyzed and four PFCs (PFOA, PFNA, PFDA and PFHpA) were found at low levels (<1 ng/L). Because of higher levels found in drinking water at several locations, some provisional drinking water guideline values for PFOS and PFOA have already been established, e.g., in the UK, Bavaria, and Minnesota. Since PFCs are present both in food and drinking water, Tolerable Daily Intake values for PFOS and PFOA have also been proposed by several institutes in Europe and in the USA.

The ingestion of dust through hand-to-mouth transfer from indoor house dust can also be a potential source of PFC exposure, especially for toddlers and children. In publications on PFCs in indoor dust, the mean PFOS and PFOA levels varied between 39 and 1,200 ng/g and between 11 and 220 ng/g, respectively.

Overall, it is clear that there is still a lack of PFC exposure data for food and beverages, which renders the assessment of the contribution of the diet to total human PFC exposure uncertain. It is, therefore, appropriate that several scientific projects have recently been launched that addresses the assessment of human exposure to PFCs and related compounds from dietary sources.

References

3 M (2001) Environmental monitoring—multi-city study, water, sludge, sediment, POTW effluent and landfill leachate samples. Executive summary, 3 M Environmental Laboratory, St Paul, MN, June 25, 2001. Available from http://www.ewg.org/files/multicity_full.pdf

Barber JL, Berger U, Chaemfa C, Huber S, Jahnke A, Temme C, Jones KC (2007) Analysis of per- and polyfluorinated alkyl substances in air samples from Northwest Europe. J Environ Monit 9:530–541

Begley TH, White K, Honigfort P, Twaroski ML, Neches R, Walker RA (2005) Perfluorochemicals: potential sources of and migration from food packing. Food Add Contam 22:1023–1031

BfR (Bundesinstitut für Risikobewertung) (2006) Hohe Gehalte an perfluorierten organischen Tensiden (PFT) in Fischen sind gesundheitlich nicht unbedenklich Stellungnahme Nr.035/2006 des BfR vom 27. Juli 2006. In German only. Available at http://www.bfr.bund.de/cm/208/hohe_gehalte_an_perfluorierten_organischen_tensiden_in_fischen_sind_gesundheitlich_nicht_unbedenklich.pdf

Björklund JA, Thuresson K, de Wit CA (2009) Perfluoroalkyl compounds (PFCs) in indoor dust: concentrations, human exposure estimates, and sources. Environ Sci Technol 43:2276–2281

Boulanger B, Peck AM, Schnoor JL, Hornbuckle KC (2005) Mass budget of perfluorooctane surfactants in Lake Ontario. Environ Sci Technol 39:74–79

COT (Committee on Toxicity of Chemicals in Food, Consumer Products and the Environment) (2006a) COT statement on the tolerable daily intake for perfluorooctane sulfonate. COT statement 2006/09, October 2006. Available online at http://www.food.gov.uk/multimedia/pdfs/cotstatementpfos200609.pdf

COT (Committee on Toxicity of Chemicals in Food, Consumer Products and the Environment) (2006b) COT statement on the tolerable daily intake for perfluorooctanoic acid. COT statement 2006/10, October 2006 available online http://www.food.gov.uk/multimedia/pdfs/cotstatementpfoa200610.pdf

Cunha I, Hoff P, Van de Vijver K, Guilhermino L, Esmans E, De Coen W (2005) Baseline study of perfluorooctane sulfonate occurrence in mussels, *Mytilus galloprovincialis*, from north-central Portuguese estuaries. Mar Pollut Bull 50:1121–1145

de Voogt P, Berger U, de Coen W, de Wolf W, Heimstad E, Mclachlan M, van Leeuwen S, van Roon A (2006) PERFORCE: perfluorinated organic compounds in the European environment. Final report to the European Commission, project NEST-508967, University of Amsterdam, Amsterdam, pp 1–126

de Vos MG, Huijbregts MAJ, van den Heuvel-Greve MJ, Vethaak AD, Van de Vijver KI, Leonards PEG, van Leeuwen SPJ, Voogt P, Hendriks AJ (2008) Accumulation of perfluorooctane sulfonate (PFOS) in the food chain of the Western Scheldt estuary. Comparing field measurements with kinetic modeling. Chemosphere 70:1766–1773

Del Gobbo L, Tittlemier S, Diamond M, Pepper K, Tague B, Yeudall F, Vanderlinden L (2008) Cooking decreases observed perfluorinated compound concentrations in fish. J Agric Food Chem 56:7551–7559

EFSA (European Food Safety Authority) (2008) Opinion of the scientific panel on contaminants in the food chain on perfluorooctane sulfonate (PFOS), perfluorooctanoic acid (PFOA) and their salts, EFSA, Parma. EFSA J 653:1–131

Ericson I, Marti-Cid R, Nadal M, van Bavel B, Lindstrom G, Domingo JL (2008a) Human exposure to perfluorinated chemicals through the diet: intake of perfluorinated compounds in foods from the Catalan (Spain) Market. J Agric Food Chem 56:1787–1794

Ericson I, Nadal M, van Bavel B, Lindström G, Domingo JL (2008b) Levels of perfluorochemicals in water samples from Catalonia, Spain: is drinking water a significant contribution to human exposure? Environ Sci Pollut Res 15:614–619

EU (European Union) (2006) Directive 2006/122/EC of the European Parliament and of the Council. Official Journal of the European Union L372, 32–34, 27-12-2006

Falandysz J, Taniyasu S, Gulkowska A, Yamashita N, Schulte-Oehlmann U (2006) Is fish a major source of fluorinated surfactants and repellents in humans living on the Baltic Coast? Environ Sci Technol 40:748–751

Fromme H, Albrecht M, Angere J, Drexler H, Gruber L, Schlummer M, Parlar H, Körner W, Wanner A, Heitmann D, Roscher E, Bolte G (2007a) Integrated Exposure Assessment Survey (INES) exposure to persistent and bioaccumulative chemicals in Bavaria, Germany. Int J Hyg Environ Health 210:345–349

Fromme H, Nitschke L, Kiranoglu M, Albrecht M, Völkel W (2008) Perfluorinated substances in house dust in Bavaria, Germany. Organohalogen Compd (70:001048–001050)

Fromme H, Schlummer M, Möller A, Gruber L, Wolz G, Ungewiß J, Böhmer S, Dekant W, Mayer R, Liebl B, Twardella D (2007b) Exposure of an adult population to perfluorinated substances using duplicate diet portions and biomonitoring data. Environ Sci Technol 41:7928–7933

Fromme H, Tittlemier SA, Völkel W, Wilhelm M, Twardella D (2009) Perfluorinated compounds – exposure assessment for the general population in Western countries. Int J Hyg Environ Health 212:239–270

FSA (Food Standards Agency) (2006) Fluorinated chemicals: UK dietary intakes. Food Survey Information Sheet 11/06, London, UK

Furdui VI, Stock NL, Ellis DA, Butt CM, Whittle DM, Crozier PW, Reiner EJ, Muir DCG, Mabury SA (2007) Spatial distribution of perfluoroalkyl contaminants in lake trout from The Great Lakes. Environ Sci Technol 41:1554–1559

Giesy JP, Kannan K (2001) Global distribution of perfluorooctane sulfonate in wildlife. Environ Sci Technol 35:1339–1342

Goosey E, Abou-Elwafa AM, Harrad S (2008) Dust from primary school and nursery classrooms in the UK: its significance as a pathway to exposure to PFOS, PFOA, HBCDs and TBBP-A. Organohalogen Compd 70:855–858

Gruber L, Schlummer M, Ungewiss J, Wolz G, Moeller A, Weise N, Sengl M, Frey S, Gerst M, Schwaiger J (2007) Tissue distribution of perfluorooctanesulfonate (PFOS) and perfluorooctanoic acid (PFOA) in fish. Organohalogen Compd 69:3

Gulkowska A, Jiang Q, So MK, Taniyasu S, Lam PKS, Yamashita N (2006) Persistent perfluorinated acids in seafood collected from two cities of China. Environ Sci Technol 40:3736–3741

Harada K, Saito N, Sasaki K, Inoue K, Koizumi A (2003) Perfluorooctane sulfonate contamination of drinking water in the Tama River, Japan: estimated effects on resident serum levels. Bull Environ Contam Toxicol 71:31–36

Hart K, Kannan K, Tao L, Takahashi S, Tanabe S (2008) Skipjack tuna as a bioindicator of contamination by perfluorinated compounds in the ocean. Sci Total Environ 403:215–221

Haukås M, Berger U, Hop H, Gulliksen B, Gabrielsen GW (2007) Bioaccumulation of per- and polyfluorinated alkyl substances in selected species from the Barents Sea food web. Environ Pollut 148:360–371

Hoff PT, Van de Vijver KI, Van Dongen W, Esmans EL, Blust R, De Coen W (2003) Perfluorooctane sulfonic acid (PFOS) in bib (*Trisopterus luscus*) and plaice (*Pleuronectes platessa*) from the Western Scheldt and the Belgian North Sea: distribution and effect. Environ Toxicol Chem 22:608–614

Hölzer J, Midasch O, Rauchfuss K, Kraft M, Reupert R, Angerer J, Kleeschulte P, Marschall N, Wilhelm M (2008) Biomonitoring of perfluorinated compounds in children and adults exposed to perfluorooctanoate (PFOA)-contaminated drinking water. Environ Health Perspect 116: 651–657

Houde M, Bujas TAD, Small J, Wells RS, Fair PA, Bossart GD, Solomon KR, Muir DCG (2006) Biomagnification of perfluoroalkyl compounds in the bottlenose dolphin (*Tursiops truncatus*) food web. Environ Sci Technol 40:4138–4144

HPA (Health Protection Agency) (2009) Maximum acceptable concentrations of perfluorooctane sulfonate (PFOS) and perfluorooctanoic acid (PFOA) in drinking water. Available online http://www.hpa.nhs.uk/webw/HPAweb&HPAwebStandard/HPAweb_C/1195733828490?p= 1158313435037

Kannan K, Corsolini S, Falandysz J, Fillmann G, Kumar KS, Loganathan BG, Mohd MA, Olivero J, Van Wouwe N, Yang JH, Aldous KM (2004) Perfluorooctanesulfonate and related fluorochemicals in human blood from several countries. Environ Sci Technol 38:4489–4495

Kannan K, Hansen JH, Wade TL, Giesy GP (2002) Perfluorooctane sulfonate in oysters, *Crassostrea virginica*, from the Gulf of Mexico and the Chesapeake Bay, USA. Arch Environ Contam Toxicol 42:313–318

Kannan K, Tao L, Sinclair E, Pastva SD, Jude DJ, Giesy JP (2005) Perfluorinated compounds in aquatic organisms at various trophic levels in a Great Lakes food chain. Arch Environ Contam Toxicol 48:559–566

Kärrman A, Ericson I, van Bavel B, Darnerud PO, Aune M, Glynn A, Lignell S, Lindström G (2007) Exposure of perfluorinated chemicals through lactation: levels of matched human milk and serum and a temporal trend, 1996–2004, in Sweden. Environ Health Perspect 115: 226–230

Kärrman A, Harada KH, Inoue K, Takasuga T, Ohi E, Koizumi A (2009) Relationship between dietary exposure and serum perfluorochemical (PFC) levels – A case study. Environ Int 35: 712–717

Kidd KA, Bootsma HA, Hesslein RH (2001) Biomagnification of DDT through the benthic and pelagic food webs of Lake Malawi, East Africa: importance of trophic level and carbon source. Environ Sci Technol 35:14–20

Kubwabo C, Stewart B, Zhu J, Marro L (2005) Occurrence of perfluorosulfonates and other perfluorochemicals in dust from selected homes in the city of Ottawa, Canada. J Environ Monit 7:1074–1078

Loos R, Wollgast J, Huber T, Hanke G (2007) Polar herbicides, pharmaceutical products, perflurooctanesulfonate (PFOS), perfluorooctanoate (PFOA), and nonylphenol and its carboxylates and ethoxylates in surface and tap waters around lake Maggiore in Northern Italy. Anal Bioanal Chem 387:1469–1478

Martin JW, Mabury SA, Solomon K, Muir DCG (2003a) Bioconcentration and tissue distribution of perfluorinated acids in rainbow trout (*Oncorhynchus mykiss*). Environ Toxicol Chem 22:196–204

Martin JW, Mabury SA, Solomon K, Muir DCG (2003b) Dietary accumulation of perfluorinated acids in juvenile rainbow trout (*Oncorhynchus mykiss*). Environ Toxicol Chem 22:189–195

Martin JW, Muir DCG, Moody CA, Ellis DA, Kwan WC, Solomon KR, Mabury SA (2002) Collection of airborne fluorinated organics and analysis by gas chromatography/chemical ionization mass spectrometry. Anal Chem 74:584–590

Martin JW, Smithwick MM, Braune BM, Hoekstra PF, Muir DCG, Mabury SA (2004a) Identification of long-chain perfluorinated acids in biota from the Canadian Arctic. Environ Sci Technol 38:373–380

Martin JW, Whittle DM, Muir DCG, Mabury SA (2004b) Perfluoroalkyl contaminants in a food web from Lake Ontario. Environ Sci Technol 38:5379–5385

Minnesota Department of Health (2007) Fish consumption advisory program meal advice categories based on levels of PFOS in fish. http://wwwhelathstatmnus/divs/eh/fish/eating/mealadivcetable4pdf

Mons M, van Roon A, de Voogt P (2007) Perfluoroalkylated substances in Dutch drinking water sources, KIWA water research, BTO 07.048, Nieuwegein, 2007

Moriwaki H, Takata Y, Arakawa R (2003) Concentrations of perfluorooctane sulfonate (PFOS) and perfluorooctanoic (PFOA) in vacuum cleaner dust collected in Japanese homes. J Environ Monit 5:753–757

Nakata H, Kannan K, Nasu T, Cho H-S, Sinclair E, Takemura A (2006) Perfluorinated contaminants in sediments and aquatic organisms collected from shallow water and tidal flat areas of the Ariake Sea, Japan: environmental fate of perfluorooctane sulfonate in aquatic ecosystems. Environ Sci Technol 40:4916–4921

Nakata A, Katsumata T, Iwasaki Y, Ito R, Saito K, Izumi S, Makino T, Kishi R, Nakazawa H (2007) Measurement of perfluorinated compounds in human milk and house dust. Organohalogen Compd 69:2844–2846

Nania V, Pellegrini GE, Fabrizi L, Sesta G, De Sanctis P, Lucchetti D, Di Pasquale M, Coni E (2009) Monitoring of perfluorinated compounds in edible fish from the Mediterranean Sea. Food Chem 115:951–957

New Jersey Department of Environmental Protection Division of Water Supply (2007) Determination of perfluorooctanoic acid (PFOA) in aqueous samples, January 2007

Olsen GW, Burris JM, Ehresman DJ, Froehlich JW, Seacat AM, Butenhoff JL, Zobel LR (2007) Half-life of serum elimination of perfluorooctanesulfonate, perfluorohexanesulfonate, and perfluorooctanoate in retired fluorochemical production workers. Environ Health Perspect 115:1298–1305

Saito N, Harada K, Inoue K, Sasaki Y, Yoshinaga T, Koizumi A (2004) Perfluorooctanoate and perfluorooctane sulfonate concentrations in surface water in Japan. J Occup Health 46: 49–59

Schrap SM, Pijnenburg AMCM, Geerdink RB (2004) Geperfluoreerde verbindingen in Nederlands oppervlaktewater; een screening in 2003 van PFOS en PFOA RIKZ rapport 2004037 Rijkswaterstaat/Rijksinstituut voor Kust en Zee, Den Haag (in Dutch only)

Schriks M, Heringa MC, van der Kooi MME, de Voogt P, van Wezel AP (2010) Toxicological relevance of emerging contaminants for drinking water quality. Water Res 44:461–476

Shoeib M, Harner T, Wilford BH, Jones KC, Zhu J (2005) Perfluorinated sulfonamides in indoor and outdoor air and indoor dust: occurrence, partitioning, and human exposure. Environ Sci Technol 39:6599–6606

Skutlarek D, Exner M, Färber H (2006) Perfluorinated surfactants in surface and drinking waters. Environ Sci Pollut Res 13:299–307

Stock NL, Lau FK, Ellis DA, Martin JW, Muir DCG, Marbury SA (2004) Polyfluorinated telomere alcohols and sulfonamides in the North American troposphere. Environ Sci Technol 38: 991–996

Strynar MJ, Lindstrom AB (2008) Perfluorinated compounds in house dust from Ohio and North Carolina, USA. Environ Sci Technol 42:3751–3756

Takagi S, Adachi F, Miyana K, Koizumi Y, Tanaka H, Mimura M, Watanabe I, Tanabe S, Kannan K (2008) Perfluorooctanesulfonate and perfluorooctanoate in raw and treated tap water from Osaka, Japan. Chemosphere 72:1409–1412

Taniyasu S, Kannan K, Horii Y, Horri Y, Hanari N, Yamashita N (2003) A survey of perfluorooctane sulfonate and related perfluorinated organic compounds in water, fish, birds, and humans from Japan. Environ Sci Technol 37:2634–2639

Tittlemier SA, Pepper K, Edwards L (2006) Concentrations of perfluorooctanesulfonamides in Canadian Total Diet Study composite food samples collected between 1992–2004. J Agric Food Chem 54:8385–8389

Tittlemier SA, Pepper K, Seymour C, Moisey J, Bronson R, Cao X-L, Dabeka RW (2007) Dietary exposure of Canadians to perfluorinated carboxylates and perfluorooctane sulfonate via consumption of meat, fish, fast foods, and fast food items prepared in their packaging. J Agric Food Chem 55:3203–3210

Tomy GT, Budakowski W, Halldorson T, Helm PA, Stern GA, Friesen K, Pepper K, Tittlemier SA, Fisk AT (2004b) Fluorinated organic compounds in an Eastern Arctic marine food web. Environ Sci Technol 38:6475–6481

Tomy GT, Tittlemier SA, Palace VP, Budakowski WR, Brarkevelt E, Brinkworth L, Friesen (2004a) Biotransformation of N-ethyl perfluorooctanesulfonamide by rainbow trout (*Oncorhynchus mykiss*) liver microsomes. Environ Sci Technol 38:758–762

Trudel D, Horowitz L, Wormuth M, Scheringer M, Cousins IT, Hungerbuehler K (2008) Estimating consumer exposure to PFOS and PFOA. Risk Anal 40:251–269

US EPA (Environmental Protection Agency) (2001) Analysis of PFOS, FOSA, and PFOA from various food matrices using HPLC electrospray/mass spectrometry. 3 M study conducted by Centre Analytical Laboratories, Inc. http://www.ewg.org/files/multicity_-full.pdf. Accessed 4 Jan 2008

Van de Vijver IK, Hoff PT, Van Dongen W, Esmans E, Blust R, De Coen W (2003) Exposure patterns of perfluorooctane sulfonate in aquatic invertebrates from the Western Scheldt estuary and the southern North Sea. Environ Toxicol Chem 22:2037–2041

Van den Heuvel-Greve M, Leonards P, Vethaak D (2006) Dioxin research Westerschelde: measuring percentages of dioxins, dioxin-like substances and other possible problematic substances in fishery products, sediment and food chains of the Westerschelde [Dioxineonderzoek Westerschelde: meting van gehalten aan dioxinen, dioxineachtige stoffen en andere mogelijke probleemstoffen in visserijproducten, sediment en voedselketens van de Westerschelde] Rapport RIKZ, 2006011 RIKZ: Middelburg, The Netherlands 80 pp

Vestergren R, Cousins IT, Trudel D, Wormuth M, Shseringer M (2008) Estimating the contribution of precursor compounds in consumer exposure to PFOS and PFOA. Chemosphere 73:1617–1624

Xu L, Krenitsky DM, Seacat AM, Butenhoff JL, Anders MW (2004) Biotransformation of N-ethyl-N-(2-hydroxyethyl)perfluorooctanesulfonamide by rat liver microsomes, cytosol, and slices and by expressed rat and human cytochromes P450. Chem Res Toxicol 17:767–775

Ye X, Strynar MJ, Nakayama SF, Varns J, Helfant L, Lazorchak J, Lindstrom AB (2008) Perfluorinated compounds in whole fish homogenates from the Ohio, Missouri, and Upper Mississippi rivers, USA. Environ Pollut 156:1227–1232

Index

P. de Voogt (ed.), *Reviews of Environmental Contamination and Toxicology*,
Reviews of Environmental Contamination and Toxicology 208,
DOI 10.1007/978-1-4419-6880-7, © Springer Science+Business Media, LLC 2010